THERE'S SOMETHING
ABOUT GÖDEL

THERE'S SOMETHING ABOUT

GÖDEL

THE COMPLETE GUIDE TO THE INCOMPLETENESS THEOREM

FRANCESCO BERTO

WILEY-BLACKWELL

A John Wiley & Sons, Ltd., Publication

This edition first published in English 2009
English translation © 2009 Francesco Berto

Original Italian text (*Tutti pazzi per Gödel!*) © 2008, Gius. Laterza & Figli, All rights reserved
Published by agreement with Marco Vigevani Agenzia Letteraria

Edition history: Gius. Laterza & Figli (1e in Italian, 2008); Blackwell Publishing Ltd (1e in English, 2009)

Blackwell Publishing was acquired by John Wiley & Sons in February 2007. Blackwell's publishing program has been merged with Wiley's global Scientific, Technical, and Medical business to form Wiley-Blackwell.

Registered Office
John Wiley & Sons Ltd, The Atrium, Southern Gate, Chichester, West Sussex, PO19 8SQ, United Kingdom

Editorial Offices
350 Main Street, Malden, MA 02148-5020, USA
9600 Garsington Road, Oxford, OX4 2DQ, UK
The Atrium, Southern Gate, Chichester, West Sussex, PO19 8SQ, UK

For details of our global editorial offices, for customer services, and for information about how to apply for permission to reuse the copyright material in this book please see our website at www.wiley.com/wiley-blackwell.

The right of Francesco Berto to be identified as the author of this work has been asserted in accordance with the Copyright, Designs and Patents Act 1988.

Library of Congress Cataloging-in-Publication Data

Berto, Francesco.
　[Tutti pazzi per Gödel! English]
　There's something about Gödel! : the complete guide to the incompleteness theorem / Francesco Berto.
　　p.　cm.
　Includes bibliographical references and index.
　ISBN 978-1-4051-9766-3 (hardcover : alk. paper) — ISBN 978-1-4051-9767-0 (pbk. : alk. paper) 1. Incompleteness theorem. 2. Gödel's theorem. 3. Mathematics–Philosophy. 4. Gödel, Kurt. I. Title.
QA9.54B4713 2009
511.3–dc22
　　　　　　　　　　　　　　　　　　　2009020156

A catalogue record for this book is available from the British Library.

Set in 10.5/13pt Garamond by SPi Publisher Services, Pondicherry, India

1　2009

Contents

For Marta Rossi

Prologue

In 1930, a youngster of about 23 proved a theorem in mathematical logic. His result was published the following year in an Austrian scientific review. The title of the paper (written in German) containing the proof, translated, was: "On Formally Undecidable Propositions of *Principia mathematica* and Related Systems I." *Principia mathematica* is a big three-volume book, written by the famous philosopher Bertrand Russell and by the mathematician Alfred North Whitehead, and including a system of logical-mathematical axioms within which all mathematics was believed to be expressible and provable. The theorem proved by the youngster referred to (a modification of) that system. It is known to the world as the Incompleteness Theorem, and its proof is one of the most astonishing argumentations in the history of human thought. The unknown youngster's name was Kurt Gödel, and the book you are now holding in your hands is a guide to his Theorem.

In fact, in his paper Gödel presented a sequence of theorems, but the most important among them are Theorem VI, and the last of the series, Theorem XI. These are nowadays called, respectively, Gödel's First and Second Incompleteness Theorems. When scholars simply talk of Gödel's Incompleteness Theorem, they usually refer to the conjunction of the two.

Gödel's Theorem is a technical result. Its original proof included such innovative techniques that in 1931 (and for years to follow) many logicians, philosophers, and mathematicians of the time – from Ernst Zermelo to Rudolf Carnap and Russell himself – had a hard time understanding exactly what had been accomplished. Nowadays, (the proof of) the Theorem is not considered too complex, and all logicians have met it, in some version or other, in some textbook of intermediate logic. Nevertheless, it remains a technical fact, inaccessible to amateurs. It is therefore surprising how much this proof has changed our understanding of logic, perhaps of mathematics and, according to some, even of ourselves and our world.

Everyone agrees, to begin with, that Gödel's result is a *terrific* achievement. Gödel's official biographer John Dawson has noted that it seems customary to invoke geological metaphors in this context. Here is Karl Popper:

> The work on formally undecidable propositions was felt as an earthquake, particularly also by Carnap.[1]

And here is John von Neumann, Princeton's "human calculator," in a speech he gave in 1951 when Gödel was given the Einstein Award:

> Kurt Gödel's achievement in modern logic is singular and monumental – indeed it is more than a monument, it is a landmark which will remain visible far in space and time.[2]

As for the legendary friendship between Gödel and Einstein, the latter once confessed to the economist Oskar Morgenstern that he had gone to Princeton's Institute for Advanced Study just "um das Privileg zu haben, mit Gödel zu Fuss nach Hause gehen zu dürfen" – to have the privilege of walking home with Gödel.

But this is not enough. Other technical results in contemporary mathematics have received attention from popular books and newspapers. Recently, this happened with Andrew Wiles' proof of Fermat's Last Theorem (a 130 page demonstration – in fact, a proof of the Taniyama–Shimura conjecture on elliptic curves, which in its turn entails Fermat's Theorem) that has inspired a nice book by Simon Singh.[3] However, no mathematical result has ever had extra-mathematical

---

[1] Quoted in Dawson (1984), p. 74.

[2] *The New York Times*, March 15, 1951, p. 31.

[3] Fermat's Last Theorem (which before Wiles' proof should rather have been called Fermat's *conjecture*) says that no equation of the simple form $x^n + y^n = z^n$ has solutions in positive integers for n greater than 2. Pierre de Fermat became famous because he claimed he had a "marvelous proof" of this fact, which unfortunately the page margin of the book on Diophantine equations he was reading was too narrow to contain.

outcomes remotely comparable to Gödel's Theorem. Speaking of geological metaphors, let us listen to Rebecca Goldstein:

> This man's theorem is the third leg, together with Heisenberg's uncertainty principle and Einstein's relativity, of that tripod of theoretical cataclysms that have been felt to force disturbances deep down in the foundations of the "exact sciences." The three discoveries appear to deliver us into an unfamiliar world, one so at odds with our previous assumptions and intuitions that, nearly a century on, we are struggling to make out where, exactly, we have landed.[4]

Gödel's Theorem has been taken as an icon of contemporary culture: a culture, it is claimed, dominated by such things as relativism, postmodernism, the twilight of incontrovertible truth, objectivity, and so on. Such words as "indeterminacy" and "incompleteness," of course, evoke this dominant thought of our time. So it has been claimed that "Gödel's incompleteness theorem shows that it is not possible to prove that an objective reality exists," or that "Religious people claim that all answers are found in the Bible [but] Gödel seems to indicate it cannot be true."[5] Gödel's result has been referred to in lots of science fiction stories, such as Christopher Cherniak's *The Riddle of the Universe*, Rudy Rucker's *Software*, and Stanislaw Lem's *Golem XIV*. Hans Magnus Enzensberger has dedicated to it the poem "Hommage à Gödel," on which Hans Werner Henze has based a violin concerto. Various bigwigs of contemporary thought have felt the need to say something on the Theorem: from Wittgenstein's very controversial claims in the *Remarks on the Foundations of Mathematics*; to Roger Penrose, who in his famous *The Emperor's New Mind* has resorted to the Theorem to defend, against the basic tenets of artificial intelligence, the idea that the human mind will never be emulated by a computer; to Douglas Hofstadter, who, with his *Gödel, Escher, Bach: An Eternal Golden Braid*, has captured a generation and has won, among other things, the Pulitzer prize.

Besides explaining what Gödel's Theorem is, in this book I aim at saying something on the extra-mathematical phenomenon it has originated. I would like, if not myself to answer, to put you in a position to answer

---

[4] Goldstein (2005), pp. 21–2.
[5] Quoted in Franzén (2005), p. 2.

the question: why are we all crazy for Gödel? How has it happened that such a hieroglyph as

> *For every ω-consistent recursive class κ of* FORMULAS *there are recursive* CLASS SIGNS *r such that neither v* Gen *r nor* Neg(*v* Gen *r*) *belongs to Flg(κ)* (where *v* is the FREE VARIABLE of *r*)[6]

has become one of the *emblems* of our culture?

Therefore, the book is divided into two parts. The first part, whose title is "The Gödelian Symphony," is the real introduction to the Theorem. The title takes seriously Ernest Nagel and James R. Newman, who, in their classic book *Gödel's Proof*, have claimed that such a proof is an "amazing intellectual symphony."[7] After a quick introduction to the historical context in which Gödel's breakthrough took place (from Frege and Russell's foundationalist project, to the discovery of the paradoxes of set theory, to the advent of Hilbert's Program), Gödel's proof is reconstructed bit by bit through an explanation of each of its key steps in a separate chapter. The details of the proof, as I have said, are intricate, but the overall strategy isn't; on the contrary, it is based on a couple of beautifully simple ideas.

The second part of the book, called "The World after Gödel," is for the most part less technical, and takes into account some of the most famous theses based upon, or allegedly following from, the Incompleteness Theorem in metaphysics, the philosophy of mathematics, the philosophy of mind, even sociology and politics. Here I show how some interpretations and uses of the Theorem are red herrings based on curious misunderstandings, whereas others are quite interesting and testify to its extraordinary fruitfulness. It is here that the aforementioned big guys, such as Wittgenstein and Penrose, come into play.

This second part is thoroughly and unashamedly influenced by *Gödel's Theorem: An Incomplete Guide to Its Use and Abuse* – a 2005 book by Torkel Franzén. In the extensive literature on Gödel, Franzén's book stands out as a serious, moderate but severe analysis of the arbitrary exploitations of Gödel's results within postmodern thought, political theory, art, religion, new age philosophy, and so on. In these

---

[6] Which is Gödel's original formulation of the First Incompleteness Theorem (see Gödel (1931), p. 30).

[7] Nagel and Newman (1958), p. 104.

contexts, the Theorem is often inaccurately quoted, misunderstood, and put to work in order to "prove" more or less anything. Among the metaphors targeted by Franzén are some of those proposed by Hofstadter – whose metaphysical journeys between Escher's paradoxical images, Bach's compositions, Zen Buddhism, the issues of artificial intelligence, etc., have been criticized by various experts in mathematical logic:

> Finding suggestions, metaphors, and analogies in other fields when studying the human mind is of course perfectly legitimate and may be quite useful. But it can only be a starting point, and actual theories and studies of the human mind would be needed to give substance to reflections like Hofstadter's. Metaphorical invocations of Gödel's theorem often suffer from the weakness of giving such satisfaction to the human mind that they tend to be mistaken for incisive and illuminating observations.[8]

Now my heart lies with Hofstadter, to whom I owe (together with millions of people) my interest in Gödel. I believe Hofstadter has put in *Gödel, Escher, Bach* all the logical and mathematical competence one can expect from a book on the Incompleteness Theorem. His only mistake, perhaps, is to have put there even *more* – specifically, enough strange stuff to make the volume unpalatable to some scholars who take themselves too seriously. I also believe that Franzén's book is a treasure-trove, and the second part of *this* book is mostly consonant with (and indebted to) his guide to the use and abuse of Gödel's Theorem. In particular, it is consonant with its underlying motivation: few cultural attitudes are as blameworthy as the instrumental usage of a product of human genius, in order to make it say whatever one wants it to.

To a certain extent, it's a question of approach. I'd reverse Franzén's claims of the kind I have just quoted – "This may be interesting, but it's just an analogy and doesn't follow logically from the Theorem" – into "This may not follow from the Theorem, but it's an interesting analogy." With a slight divergence of attitude, those who write textbooks on Gödel (and there are so many of them) can make the difference between a world in which readers get honestly interested in his Theorem, and

---

[8] Franzén (2005), pp. 124–5.

one in which people come to believe that, all in all, logic has nothing important to tell us, for it is too difficult and esoteric a subject to matter in our lives.

* * *

What should you already know in order to understand this book? Basically, you need to have attended a course of elementary logic, and/ or to have read an introduction to elementary logic (lots of good textbooks are available). Actually, the book presupposes even *less*: it presupposes, roughly, the intersection of the material covered in the standard courses and the handbooks for dummies. For instance, some logic courses explain the basics of set theory, some don't. To make thing easier, therefore, I have introduced in the first chapter the basic set-theoretic machinery we shall need. Furthermore, no specific mathematical competence is required. However, you are expected to have some idea of what a logical operator is, for if I had to explain these notions as well, the book would have turned out to be far too long. In any case, most of the (not too numerous) logical formulas can be skipped without any substantial loss, since I have usually provided the corresponding translations into ordinary English.

I have tried to explain things in as friendly a way as possible: as a philosopher, I believe one should never dispense with comprehensibility for the sake of exactness. One consequence of this is a certain laxity with the use-mention distinction: Quine's quasi-quotation marks, as well as other quotation devices, have been avoided in all cases in which no confusion would have resulted. Formal as well as informal expressions have been used as names of themselves, disambiguation being secured by the context.

This brings me to say something about the reasons I wrote this book. The first is the same that led me to write others: in the long run, I want to become rich and famous.

The second reason is that, even though I happen to give undergraduate and postgraduate courses in logic, I am not a mathematician at all, but a philosopher who often encounters logic's sharpness and incredible levels of accuracy and abstractness. After some years of interest in the discipline, I sometimes still have a hard time when learning new and advanced things. I also believe that most people out there, when it's about getting in touch with such difficult subjects as mathematical logic, work more or less like me. And this book should resemble as much as possible the book *I* would have liked to have had in my hands when I began to learn about Gödel's Theorem: a text which would explain things to me without oversimplifying, but also without drowning me in long sequences of formal proofs that my philosophical mind would have had trouble following; a book which would take me by the hand and guide me in my ascent towards the Gödelian peak. If the resulting work in some measure approaches this, I will have achieved one of my goals.

Finally, in case you are wondering, I know why *I* am crazy for Gödel: because following his amazing proof has been one of my most fascinating intellectual experiences. Should I succeed in imparting some of this Gödelian magic, I will have achieved another of my goals.

Acknowledgments

This is the English version of a book I have published in Italian with the title *Tutti pazzi per Gödel!* (Laterza, Roma, 2007), and the number of people I am to credit has, understandably, increased for this new edition.

I am grateful to Nick Bellorini of Wiley-Blackwell and to Anna Gialluca of Laterza for their commitment and support during the whole editorial process.

My general debts to those who supported me in writing (and rewriting) the book range across three continents. To begin with, I have some French debts to Friederike Moltmann, Jacques Dubucs, Alexandra Arapinis, and the members of the Institute of History and Philosophy of Sciences and Techniques of the Sorbonne University (Paris 1/CNRS/ ENS) for hosting me in their prestigious research center. During my "Chaire d'excellence" fellowship in Paris, they have provided me with a very comfortable environment to carry out my work; I hope this book partly repays them for their trust (or at the very least for the hundreds of *cafés avec biscuits* I have guzzled in my bureau at the École Normale Superiéure). Thanks also to my wonderful Parisian friends Valeria, Carlo, and Giulia for great conversations that helped make the life of an Italian émigré more enjoyable.

Next, I have some English debts to Dov Gabbay, John Woods, and Jane Spurr of King's College London for their proposal to publish my *How to Sell a Contradiction* (College Publications, London, 2007), and for kind permission to reuse material taken from Chapters 1, 4, and 12 of that book throughout this one.

Australia is the place where my heart belongs. Down under, I am indebted to Graham Priest for constant support and tremendously useful comments on my work in paraconsistent logic, Meinongian ontology, and Gödelian issues. Thanks also to Paul Redding and Mark Colyvan of the University of Sydney, Greg Restall of the University of Melbourne, and Ross Brady and Andrew Brennan of La Trobe for support of the most varied kind. Part of the material included in Chapter 12 was presented at the Fourth World Congress on Paraconsistency held in Melbourne in July 2008; I am grateful to all participants and especially to J.C. Beall, Koji Tanaka, Zach Weber, Diderik Batens, and Francesco Paoli for comments, encouragement, and enjoyable discussions.

In the US I am indebted, to begin with, to Vittorio Hösle of the University of Notre Dame: during the scholarship I was offered in 2006, I proposed for the first time my Gödelian reflections in a seminar on the philosophy of mind of the twentieth century, and parts of that talk now appear in Chapter 11. I am grateful to David Leech, Gregor Damschen, Dennis Monokroussos, Dae-Joong Kwon, Miguel Perez, Ricardo Silvestre, Fernando Suàrez, and Nora Kreft for lively discussion and comments during that hot Indiana summer. Next, I am most grateful to Achille Varzi of Columbia, NY, for his wonderful support throughout these years and his very encouraging comments on the manuscript of the book.

Thanks to the professors and researchers of various Italian universities for their efforts to support me in the precarious circumstances of academic life: Vero Tarca, Luigi Perissinotto, Luca Illetterati, Max Carrara, Franco Chiereghin, Antonio Nunziante, Francesca Menegoni, Giuseppe Micheli, Andrea Tagliapietra, Michele Di Francesco, Emanuele Severino, Andrea Bottani, Richard Davies, Mauro Nasti, Vincenzo Vitiello, Massimo Adinolfi, and Franca d'Agostini.

Thanks to Enrico Moriconi and Dario Palladino for their invaluable textbook expositions of Gödel's Theorem – a secure guide both for beginners and for advanced scholars.

Finally, very special thanks to Diego Marconi, whose philosophical work has influenced me so much in so many ways; and to Marcello Frixione, the amazing Matt Plebani, and Blackwell's anonymous referee, for accurate and extensive comments to the contents of the whole book.

Part I

The Gödelian Symphony

One of themselves, even a prophet of their own, said, The Cretans are always liars … This witness is true. (St Paul, *Epistle to Titus,* 1: 12–13)

1

Foundations and Paradoxes

In this chapter and the following, we shall learn lots of things in a short time.[1] Initially, some of the things we will gain knowledge of may appear unrelated to each other, and their overall usefulness might not be clear either. However, it will turn out that they are all connected within Gödel's symphony. Most of the work of these two chapters consists in preparing the instruments in order to play the music. We will begin by acquiring familiarity with the phenomenon of *self-reference* in logic – a phenomenon which, according to many, has to be grasped if one is to understand the deep meaning of Gödel's result. Self-reference is closely connected to the famous *logical paradoxes*, whose understanding is also important to fully appreciate the Gödelian construction – a construction that, as we shall see, owes part of its timeless fascination to its getting quite close to a paradox without falling into it.

But what is a paradox? A common first definition has it that a paradox is the absurd or blatantly counter-intuitive conclusion of an argument, which starts with intuitively plausible premises and advances via seemingly acceptable inferences. In *The Ways of Paradox*, Quine claims that "a paradox is just any conclusion that at first sounds absurd but that has an argument to sustain it."[2] We shall be particularly concerned not just with sentences that are paradoxical in the sense of being implausible, or contrary to common sense ("paradox" intended as something opposed to the δόξα, or to what is ἔνδοξον, entrenched in pervasive

---

[1] This chapter draws on Berto (2006a), (2007a), and (2007b) for an account of the basics of set theory and of logical paradoxes.
[2] Quine (1966), p. 1. Sainsbury's definition is: "an apparently unacceptable conclusion derived by apparently acceptable reasoning from apparently acceptable premises" (1995), p. 1.

and/or authoritative opinions), but with sentences that constitute authentic, full-fledged contradictions. A paradox in this strict sense is also called an *antinomy*.

However, sometimes the whole argument is also called a paradox.[3] So we have Graham Priest maintaining that "[logical] paradoxes are all arguments starting with apparently analytic principles ... and proceeding via apparently valid reasoning to a conclusion of the form 'α and not-α'."[4]

Third, at times a paradox is considered as a set of jointly inconsistent sentences, which are nevertheless credible when addressed separately.[5]

The logical paradoxes are usually subdivided into the *semantic* and *set-theoretic*. What is semantics, to begin with? We can understand the notion by contrasting it with that of *syntax*. Talking quite generally, in the study of a language (be it a natural language such as English or German, or an artificial one such as the notational systems developed by formal logicians), semantics has to do with the relationship between the linguistic signs (words, noun phrases, sentences) and their meanings, the things those signs are supposed to signify or stand for. Syntax, on the other hand, has to do with the symbols themselves, with how they can be manipulated and combined to form complex expressions, without taking into account their (intended) meanings.

Typically, such notions as *truth* and *denotation* are taken as pertaining to semantics.[6] Importantly, a linguistic notion is classified as (purely)

---

[3] Beall and van Fraassen (2003), p. 119 claim that "a *paradox* ... is an argument with apparently true premises, apparently valid reasoning, and an apparently false (or untrue) conclusion."

[4] Priest (1987), p. 9.

[5] This definition is taken as having some advantages over the previous ones by Sorensen (2003), p. 364.

[6] And *truth* is generally considered as the basic semantic notion. This is because, of the various syntactic categories, the dominant paradigm of contemporary philosophy of language puts (declarative) sentences at the core, and takes the meaning of sentences to consist mainly, if not exclusively, of their truth conditions. The celebrated motto comes from Wittgenstein's *Tractatus logico-philosophicus*: "To understand a proposition means to know what is the case if it is true." Since to understand a sentence is to grasp its meaning, the motto says that this amounts to understanding the conditions under which the sentence at issue is true. To know what "Snow is white" means is to know what the world must be like if this sentence has to be true. And it is true in the event that things in the world are as it claims them to be, that is, in the event that snow is actually white. Within this semantic perspective (which is therefore called "truth-conditional"), precisely the notion of truth is placed at center stage.

syntactic when its specification or definition does not refer to the meanings of linguistic expressions, or to the truth and falsity of sentences. The distinction between syntax and semantics is of the greatest importance: I shall refer to it quite often in the following, and the examples collected throughout the book should help us understand it better and better.

The set-theoretic paradoxes concern more technical notions, such as those of *membership* and *cardinality*. These paradoxes have cast a shadow over set theory, whose essentials are due to the great nineteenth-century mathematician Georg Cantor, and which was developed by many mathematicians and logicians in the twentieth century.

Nowadays, set theory is a well-established branch of mathematics. (One should speak of set *theories*, since there are many of them; but mathematicians refer mainly to one version, that due to Ernst Zermelo and Abraham Fraenkel, to which I shall refer in the following.) But the theory has also a profound philosophical importance, mainly because of the role it has had in the development of (and the debate on) the so-called *foundations* of mathematics. Between the end of the nineteenth century and the beginning of the twentieth, the great philosophers and logicians Gottlob Frege and Bertrand Russell attempted to provide a definitive, unassailable logical and philosophical foundation for mathematical knowledge precisely by means of set theory. When Gödel published his paper, the dispute on the foundations of mathematics was quite vigorous, because of a crisis produced by the discovery of some important paradoxes in the so-called naïve formulation of set theory.

In these initial chapters, therefore, we shall learn some history and some theory. On the one hand, we will have a look at the changes that logic and mathematics were undergoing at the beginning of the twentieth century, mainly because of the paradoxes: to know something of the logical and mathematical context Gödel was living in will help us understand why the Theorem was the extraordinary breakthrough it was. But we shall also learn some basic mathematical and set-theoretical concepts. Among the most important notions we will meet in this chapter is that of *algorithm*. By means of it, we should come to understand what it means for a given set to be (intuitively) *decidable*; what it means for a given set to be (intuitively) *enumerable*; and what it means for a given function to be (intuitively) *computable*. If this list of announcements on the subjects we shall learn sounds alarming, I can only say that the initial pain will be followed by the gain of seeing these separate pieces come together in the marvelous Gödelian jigsaw.

1 "This sentence is false"

I have claimed that the semantic paradoxes can involve different semantic concepts, such as *denotation*, *definability*, etc. We shall focus only on those employing the notions of truth and falsity, which are usually grouped under the label of the *Liar*. These are the most widely discussed in the literature – those for which most tentative solutions have been proposed. They are also the most classical, having been on the philosophical market for more than 2,000 years – a fact which, by itself, says something about the difficulty of dealing with them. The ancient Greek grammarian Philetas of Cos is believed to have lost sleep and health trying to solve the Liar paradox, his epitaph claiming: "It was the Liar who made me die / And the bad nights caused thereby."

One of the most ancient versions of semantic paradox appears in St Paul's *Epistle to Titus*. Paul blames a "Cretan prophet," who was to be identified as the poet and philosopher Epimenides, and who was believed to have at one time said:

(1)   All Cretans always lie.

Actually, (1) is not a real paradox in the strict sense of a sentence which, on the basis of our *bona fide* intuitions, would entail a violation of the Law of Non-Contradiction. It is just a sentence that, on the basis of those intuitions, cannot be true. It is self-defeating for a *Cretan* to say that Cretans always lie: if this were true – that is, if it were the case that all sentences uttered by any Cretan are false – then (1), being uttered by the Cretan Epimenides, would have to be false itself, against the initial hypothesis. However, there is no contradiction yet: (1) can be just false under the (quite plausible) hypothesis that some Cretan sometimes said something true.

We are dealing with a full-fledged Liar paradox (also attributed to a Greek philosopher, and probably the greatest paradoxer of Antiquity: Eubulides) when we consider the following sentence:

(2)   (2) is false.

As we can see, (2) refers to itself, because it is no. 2 of the sentences highlighted in this chapter, and tells something of the very sentence no. 2. Also (1) refers to itself, but does it in a different way from (2). This is what makes (1) not strictly paradoxical. Sentence (1) claims that all the members of a set of sentences (those uttered by Cretans) are false. In addition, it belongs to that very set, due to its being uttered by a Cretan. Therefore (1) can be simply false, under the empirical hypothesis that some sentence uttered by a Cretan, and different from (1), is true. This is also what makes it look so odd: it is unsatisfactory that a logical paradox is avoided only via the empirical fact that some Cretan sometimes said something true.

Some form of self-reference can be detected in (almost) all paradoxes, so that the phenomenon of self-reference as such has been held responsible for the antinomies. Nevertheless, lots of self-referential sentences are harmless, in that we seem to be able to ascertain their truth value in an unproblematic way. For instance, you may easily observe that, among the following, (3) and (4) are true, whereas (5) is false:

(3)   (3) is a grammatically well-formed sentence.
(4)   (4) is a sentence contained in *There's Something About Gödel!*
(5)   (5) is a sentence printed with yellow ink.

In contrast, (2) is not harmless at all. Let us reason by cases. Suppose (2) is true: then what it says is the case, so it's false. Suppose then (2) is false. This is what it claims to be, so it's true. If we accept the Principle of Bivalence, that is, the principle according to which all sentences are either true or false, both alternatives lead to a paradox: (2) is true *and* false! To claim that something is both true and false is to produce a denial of the Law of Non-Contradiction. And this is how our *bona fide* intuitions lead us to a contradiction, via a simple reasoning by cases.

Other versions of the Liar are called *strengthened Liars*,[7] or also *revenge Liars* (whereas (2) may be called the "standard" Liar):

(6)   (6) is not true.

(7)   (7) is false or neither true nor false.

The reason why sentences such as (6) deserve the title of *strengthened* Liars is the following. Some logicians (including the best one of our times, Saul Kripke) have proposed circumventing the standard Liar (2) by dispensing with the Principle of Bivalence, that is, by admitting that some sentences can be neither true nor false, and that (2) is among them. Sentence (2) is a statement such that, if it were false, it would be true, and if it were true, it would be false. But we can avoid the contradiction by granting that (2) is neither. Such a solution has some problems with sentences such as (6), which appear to deliver a contradiction even when we dismiss Bivalence. In this case, the set of sentences is divided into three subsets: the true ones, the false ones, and those which are neither. Now we can reason by cases again with (6): either (6) is true, or it is false, or neither. If it's true, then what it says is the case, so it's not true. If it's false or neither true nor false, then it is not true. However, this is what it claims to be, so in the end it's true. Whatever option we pick, (6) turns out to be both true and untrue, and we are back to contradiction. This Liar thus gains "revenge" for its cousin (2).[8]

2   The Liar and Gödel

A sentence can refer to itself in various ways, so we can have various versions of (2). For instance:

(2a)   This sentence is false.

(2b)   I am false.

(2c)   The sentence you are reading is false.

---

[7]   As far as I know, the terminology is due to van Fraassen (1968).

[8]   Some (e.g. Graham Priest, in his classic works on dialetheism) have conjectured that *any* consistent solution to the Liar faces the same destiny: for any version of the Liar paradox which is solved by the relevant theory, one can build a revenge paradox by exploiting the very notions introduced in the theory in order to address the previous one. On these issues, see Berto (2006a), Ch. 2, and Berto (2007b), Ch. 2.

The paradox can also be produced without any immediate self-reference, but via a short-circuit of sentences. For instance:

(2d)   (2e) is true.
(2e)   (2d) is false.

This is as old as Buridan (his sophism 9: Plato saying, "What Socrates says is true"; Socrates replying, "What Plato says is false"). If what (2d) says is true, then (2e) is true. However, (2e) says that (2d) is false ... and so on: we are in a paradoxical loop.

However, it seems that self-reference is obtained in all cases by means of an unavoidable "empirical," i.e., contextual or indexical, component. In fact, in the paradoxical sentences we have examined so far, self-reference is achieved via the numbering device, or via indexical expressions such as "I", "this sentence," and so on. Only factual and contextual information tells us that the denotation of (those tokens of) such expressions is the very sentence in which they appear as the grammatical subjects. This holds for the "looped Liar": suppose (2d) is as above, but (2e) now is "Perth is in Australia." Then (2d) is just true, and no paradox is expected. But it happens also with the immediately self-referential paradoxical statements above: for instance, if I uttered (a token of) (2a) by pointing, say, at (a token of) the sentence "2 + 2 = 5" written on a blackboard, there would be no self-reference at all, for "this sentence," in the context, would refer to (the token of) "2 + 2 = 5" (and, besides, I would be claiming something true). Ditto if I uttered (2c) referring to you, while you are reading the false sentence written on the blackboard.

Because of this, some (among which the Italian mathematician Giuseppe Peano, of whom I shall talk again later) have believed that the semantic paradoxes involve some non-*logical* phenomenon: they depend on contextual, empirical factors. Frank Ramsey, to whom the distinction between semantic and set-theoretic paradoxes is usually ascribed, depicted the situation thus by referring to the list of paradoxes examined in Russell and Whitehead's *Principia mathematica*:

> Group A [i.e., antinomies no. 2, 3, and 4 of the original list of *Principia*: among them, the Russell and Burali-Forti paradoxes, which I will introduce later] consists of contradictions which, were no provision made against them, would occur in a logical or mathematical system itself. They involve

only logical or mathematical terms such as class and number, and show that there must be something wrong with our logic and mathematics. But the contradictions in Group B [i.e., antinomies no. 1, 5, 6, 7 of *Principia*: among them, the Liar] are not purely logical, and cannot be stated in logical terms alone; for they all contain some reference to thought, language, or symbolism, which are not formal but empirical terms.[9]

However, just after Ramsey had proposed the distinction, Gödel himself showed how to build, within a formal logical system, self-referential constructions with no empirical trespassers of any kind: self-referential statements whose content is as empirical and contextual as that of "2 + 2 = 4." To achieve this, Gödel used the language of mathematical logic as nobody had done before; and the apparatus he put to work is probably the most inspired aspect of the proof of the Theorem that bears his name.

Behind the Gödelian construction hide precisely the simple intuitions concerning the conundrum originated by the Liar which made the ancient Greeks lose their sleep. However, Gödel did not exploit those intuitions to engender a contradiction, via a sentence that claims of itself to be false, like (2), or untrue, like (6). He produced a sentence that walks on the edge of paradox, without falling into it. I shall talk of this mysterious Gödelian sentence at length: it is, in fact, the main character of the story I have begun to tell.

3   Language and metalanguage

The great Polish logician Alfred Tarski, and many after him, have held responsible for such semantic paradoxes as the Liars certain features of natural language, grouped under the label of "semantic closure conditions." Roughly, a semantically closed language is a language capable of talking of its own semantics, of the meanings of the expressions of the language itself. Less roughly, "a semantically closed language is one with semantic predicates, like 'true', 'false', and 'satisfies', that can be applied to the language's own sentences."[10] It is because English can mention

---

[9]   Ramsey (1931), pp. 36-7.
[10]   Kirkham (1992), p. 278.

its own expressions, and ascribe semantic properties to them, that we can have such sentences as (2) or (6): some expressions of our every-day language can somehow refer to themselves; "true" and "false" are perfectly meaningful predicates of English; and they can be applied to sentences of English.

In a Tarskian approach, the semantic paradoxes are due to a mixture of *object language* and *metalanguage*. Logicians and philosophers usually call "object language" the language we speak about, or we give a theory of, this being precisely the object of the theory. However, the theory itself will obviously be phrased in some language or other; and the language in which the theory is formulated can be labeled as a metalanguage, that is to say, a "language on a language."

That (object) language and metalanguage may be distinct is fairly clear. If you are studying a basic French grammar written in English, you will find that French figures in it mostly as the object language, whereas English is employed mainly as the metalanguage. But in our self-referential statements above, the two levels are mixed: these are English sentences talking of English sentences (specifically, of themselves). And this fusion, according to the Tarskian approach, gives rise to the paradox.

The Tarskian treatment maintains that the truth predicate cannot be *univocal*. A single surface grammar expression, "is true," has an ambiguous function for different languages, each of which is semantically open at the level of some deep logical grammar. Instead of a unique language, we would have a hierarchy, more or less with the following structure. For any ordinal n we have a language L_n, and n is the *order* of L_n. Let us begin with L_0, taken as our "basic" level language. The semantic concepts concerning L_0 cannot be expressed within L_0 itself, but must be expressed in a language, say L_1, which is its metalanguage. L_1 will contain predicates that refer to the semantic concepts of L_0 (and, in particular, by means of which we can provide a definition of *truth* for sentences of L_0: I shall come to the details of the Tarskian definition of truth in Chapter 9). However, L_1 is itself "semantically open": it cannot express its own semantic concepts. So a definition of truth for L_1 will be expressed in a language, L_2, which is the metalanguage of L_1; and so on.[11] The Tarskian solution parameterizes the semantic predicates along the hierarchy of the metalanguages: the metalinguistic

[11]   See Sainsbury (1995), pp. 118–19.

"true"and "false"are now abbreviations for"true in the object language," "false in the object language." In particular, the standard Liar turns into "This sentence is false in the object language." Its place is in the metalanguage, and it is just *false* there, not paradoxical: since metalinguistic sentences do not belong in the object language at all, the Liar does not have the property it claims to have.[12]

In point of fact, however, Tarski proposed his hierarchy as a structure for the artificial languages of formal logic, and did not claim his strategy to be applicable to natural languages (though others after him have been less restrained on this). Tarski's prudence is easily understood. First, there is no evidence that the predicate "true" performs some ambiguous function along some hidden hierarchy of languages, metalanguages, metametalanguages, etc. This makes the proposal to apply the theory to ordinary English look like a form of *revisionism*: a suggestion to the effect that ordinary English be somehow regimented. If the idea came to Tarski's mind, he certainly found it unsatisfactory.[13]

Second, a decisive difference between such a hierarchy and English is that there does not seem to be *any* metalanguage for English. This becomes manifest if we accept the principle according to which ordinary language is, so to speak, "transcendental": anything that is linguistically expressible can be expressed within ordinary language – there is no limit to it. In Tarski's own words:

A characteristic feature of colloquial language (in contrast to various scientific languages) is its universality. It would not be in harmony with the spirit of this language if in some other language a word occurred which could not be translated into it; it could be claimed that"if we can speak meaningfully about anything at all, we can also speak about it in colloquial language."[14]

---

[12] See Kirkham (1992), p. 280.

[13] "Whoever wishes, in spite of all difficulties, to pursue the semantics of colloquial language with the help of exact methods will be driven first to undertake the thankless task of a reform of this language. He will find it necessary to define its structure, to overcome the ambiguity of the terms which occur in it, and finally to split the language into a series of languages of greater and greater extent, each of which stands in the same relation to the next in which a formalized language stands to its metalanguage. It may, however, be doubted whether the language of everyday life, after being 'rationalized' in this way, would still preserve its naturalness and whether it would not rather take on the characteristic features of the formalized languages" (Tarski (1936), p. 406).

[14] Tarski (1956), p. 164.

We need not enter into subtle issues in the philosophy of language, though. Two important things to be kept in mind in following this book are (a) the idea of the distinction between (object) language and meta-language, and (b) the basic intuition behind the kind of self-reference taking place when we can see a certain linguistic expression as talking of *itself*, and as ascribing to itself some features and properties.

4   The axiomatic method, or how to get the non-obvious out of the obvious[15]

We have seen that the Liar has been around since the ancient Greeks. We also need to start from ancient Greece to understand what the *axiomatic method* is, and why this method has enjoyed an almost spotless reputation throughout Western thought. We need to refer, in fact, to Euclidean geometry – a theory we all know from elementary school, and which has been the paradigm of axiomatization for centuries.

In his *Elements of Geometry*, the Greek mathematician Euclid introduced some simple geometrical definitions (such as "A point is that which has no part"), and the celebrated five postulates, or axioms, that bear his name (for instance, the first says, "Any two points can be joined by a straight line"; the fourth says, "All right angles are congruent"). In the axiomatic approach, axioms are sentences accepted without a proof, as principles of deduction – principles from which we can infer other sentences, via purely deductive reasoning. The sentences with which the various deductive chains come to an end are called the *theorems* of Euclidean geometry. Such deductive chains are the *proofs* of Euclidean geometry; the closures of such chains, i.e., the theorems, are what are properly said to have been demonstrated, or proved, from the axioms.

Axioms, proofs, theorems: this beautifully simple and powerful pattern of knowledge has always fascinated scholars. Beginning with a small amount of what Quine would have called "ideology," that is, with a few intuitive notions and the initial postulates, Euclidean geometry delivered a large amount of theorems by means of deductive procedures which appeared to be fairly clear and rigorous. And the axioms

---

[15]   This comes from the Leibnizian motto: *spernimus obvia, ex quibus tamen non obvia sequuntur.*

were considered – keep this in mind for the following – as (manifestly) *true*. In the so-called classical conception of axiomatic systems, axioms were taken as evident, if not trivial, truths. This is exactly how we wanted them, so that we could accept them without further argumentation. The chain of deductions and inferences has to come to an end somewhere, and what better place to append it than the obvious? The proofs of Euclidean geometry being valid proofs, truth could go downstairs from the axioms to often more complex and less evident theorems. This, after all, is the fundamental virtue of valid deductive reasoning: transmitting truth from premises to conclusions.

For these reasons, numerous philosophers (from Descartes to Spinoza and Kant) took Euclidean geometry as a paradigm of rigorous knowledge. They sometimes even tried to export the model and its successful features to other compartments of science, so as to raise them to a comparable level of certitude and precision. The case of "export" we are most interested in takes the stage in the next paragraph.

5   Peano's axioms ...

A closer ancestor of Gödel's results is constituted by the amazing developments of mathematics in the nineteenth century. Some of them had to do with the so-called arithmetization of analysis, which allowed the reduction of higher parts of mathematics to elementary arithmetic, that is, to the theory of natural numbers (the positive integers, including zero: 0, 1, 2, ...). Thanks to the work of mathematicians like Weierstrass, Cantor, and Dedekind, other kinds of number were referred to rational numbers (the numbers representable as ratios of integers) and, via these, to the natural numbers.

The two mathematical results we are most interested in, however, are (a) the aforementioned theory of infinite numbers and sets due to Cantor, and (b) an axiomatic achievement: the formulation of the axioms for arithmetic due to Dedekind and Peano. While the axiomatization of geometry dated back to the ancient Greeks, an analogue account for arithmetic became available only at the end of the nineteenth century, when Dedekind provided the recursive equations for addition and multiplication (which will be met and explained in a later chapter), and immediately afterwards Peano proposed the famous axioms for arithmetic that bear his name.

Only three notions appear in Peano's axioms – three notions taken as primitive and fundamental: *zero*, (natural) *number*, and (immediate) *successor*, the (immediate) successor of a number being the one that follows it immediately in the ordering of the naturals, i.e., 1 is the successor of 0, 2 is the successor of 1, and so on. In the *Arithmetices principia, nova methodo exposita* Peano employed the three basic notions to formulate the following five principles:

(P1)   Zero is a number.

(P2)   The successor of any number is a number.

(P3)   Zero is not the successor of any number.

(P4)   Any two numbers with the same successor are the same number.

(P5)   Any property of zero that is also a property of the successor of any number having it is a property of all numbers.

Peano's fifth axiom, (P5), is usually called the (mathematical) *induction principle*. I'll come back to it repeatedly in the following (as we will see, "induction" here has little to do with inductive reasoning; on the contrary, it is a typical procedure of deductive sciences).

6   … and the unsatisfied logicists, Frege and Russell

Peano's axiomatization of arithmetic, just like Euclid's for geometry, was still considered by some scholars to be an inadequate account. They complained especially about the insufficient logical rigor in the proof chains. In fact, in his *Elements* Euclid had formulated some so-called "common notions" which looked like general rules of logical inference ("Two things identical to a third one are identical to each other," for instance). However, Euclid's language lacked the rigor of modern logical languages. To establish that a given deductive chain is valid (that is, that it will never lead us from true premises to a false conclusion), one has to look at the meanings of (some of) the words and phrases used to express it. But ordinary language expressions, as we noted when we mentioned Tarski's position on natural language, are often vague, equivocal, or both. Because of this, at least since Leibniz's *Characteristica universalis,* philosophers have been envisaging artificial, formal languages to serve as antidotes to the deficiencies of natural language, and in which rigorous science could be formulated: languages whose syntax was to be absolutely precise, and whose expressions were to have completely precise and univocal meanings.

Now, some of Euclid's proofs appeared to include notions captured neither by the explicit definitions, nor by the postulates; and they certainly adopted principles of logical inference which were not listed among the common notions. As for Peano, he had already introduced a formal notation in 1888 (in fact, one including symbols which are nowadays embedded in the canonical logical and set-theoretical notation). However, his axiomatization of arithmetic lacked a rigorous specification of the logical principles employed in the deductions from the axioms. Peano's proofs were rather informal, and the task of establishing the correctness of the deductive passages was often simply left to the reader. By contrast, since the introduction to his 1879 *Ideography* – the text whose publication is considered the founding act of modern logic – Gottlob Frege had begun to show how arithmetical claims could be proved by means of precise, explicitly stated logical rules. On the one hand, arithmetical proof sequences had to be translated into an artificial symbolic language. On the other hand, the logical rules operating in the proofs had to be made rigorously explicit. Frege provided a first precise characterization of what we nowadays call a *formal system* – a notion to which I will return again and again, and whose richness will be explored little by little as this book develops.

Once "higher" mathematics had been reduced to the natural numbers, and secure (or so it seemed) logical rules to reason on them were available, one might have gained the impression that mathematical knowledge had reached safe ground. Infinitesimals and irrational numbers had a problematic status. For a long time, mathematicians and philosophers had had qualms concerning the consistency of mathematical analysis; but everyone considered the good old integers to be reliable guys. However, Frege had a deeper ambition: that of providing a *foundation, on pure logic, for arithmetic itself.* Such an ambition was shared, between the end of the nineteenth century and the beginning of the twentieth, by Bertrand Russell. It was Russell, in fact, who brought Frege's work to the attention of a wider audience; and "logicism" was the name given to the project, precisely because of its aiming at a rigorous logical foundation of arithmetic. Both Frege and Russell believed there to be no theoretical distinction between the two domains. The notions of zero, (natural) number, and (immediate) successor, taken by Peano as primitive for arithmetic, and its fundamental principles as captured by Peano's axiomatization, were to be defined and deduced in their turn from still more fundamental and purely logical principles. Specifically, they were to be obtained precisely from the principles of set theory, which at the time was considered a limb of logic.[16]

7   Bits of set theory

To understand what the logicist program consisted of – and, most importantly, what major obstacle it stumbled upon – we need to swallow some of the medicine of set theory. What's a set, to begin with? In the first instance, a set is just a collection of objects. In the following, I will usually refer to sets by means of capital Latin letters: A, B, C, ...[17] The fundamental and primitive relation at issue in set theory is that of an object *belonging to* a set, or *being a member of* a set. This is expressed by the symbol "$\in$" (and non-membership is expressed by "$\notin$"). I will write, then, such things as "$x \in A$" ("$x \notin A$"), to mean that a given object x is (is not) a member or an element of set A, that is, it belongs (does not belong) to the set.

One can sometimes specify a set simply by providing a complete list of its elements, which is usually written thus: $\{x_1, ..., x_n\}$ is the set whose

---

[16]   Nowadays the border between logic and arithmetic is much more precise, mainly because of Gödel's Theorem. When logicians talk about "logic," with no further qualification, they usually refer to the so-called *elementary* or *first-order* logic: the predicate calculus with quantifiers and identity, familiar to anyone who has attended a course in basic logic. The so-called "higher-order" logic has quite a different status, and fuzzy borders with set theory. We shall come back to these notions and acquire more familiarity with them in the following; but we should bear in mind that the sharp distinction between elementary logic and set theory is due to their quite different features, and that such differences emerged mainly thanks to Gödel's work.

[17]   Sometimes taken as variables ("For any set A ..."), sometimes as constants, that is, as names for specific sets ("Russell's set R ...," "the set N of natural numbers ..."). Context will disambiguate.

members are indeed $x_1, ..., x_n$. For instance, one can specify the set whose sole elements are Frege, Juliette Binoche, and the city of Melbourne, thus:

{Gottlob Frege, Juliette Binoche, Melbourne}.

Notice that what is properly listed are the elements' names, not the elements themselves (similarly, when one lists the players of AC Milan, one doesn't actually put in a row Kaka, Pirlo, Shevchenko,..., but writes down their names).

The only things that matter about sets are their members. It is irrelevant in what order (the names of) the members are listed, or whether they occur more than once. This means that, for instance, the following sets:

{Juliette Binoche, Melbourne, Gottlob Frege}
{Gottlob Frege, Melbourne, Melbourne, Gottlob Frege, Juliette Binoche, Juliette Binoche}

are still the same set as the one introduced above.

When we deal with sets having an infinity of elements, such as the set of natural numbers (which I shall call "N"), we cannot in practice specify (the names of) all these elements in a list: we could never finish the job.[18] Therefore sets are often introduced via a *condition*: one specifies some feature, or shared property, or characteristic enjoyed by all and only the elements of the set at issue. A standard notation is the following:

$$\{x \mid ... x ...\},$$

to be read in English as: "The set of all the x, such that ... x ...." For instance, the set of odd numbers is $\{x \mid x$ is an odd number$\}$.

Now, Frege had based his logicist program, aimed at defining numbers via set-theoretic notions, on a version of what is nowadays called

---

[18]   Which doesn't mean that one cannot think about the elements of an infinite set, such as N, as arranged in an infinite list with a first element, a second one, etc., as we do when we write: "0, 1, 2, 3, ..." – on the contrary, the theoretical possibility of arranging the naturals in one such list will prove of great importance in the following.

"naïve set theory." This is built upon two fundamental principles (supposedly) capturing our intuitive conception of set.[19] The *Extensionality Principle* spells out the sufficient conditions for identity between sets. In the canonical notation, it goes like this:

(EP) $\forall x(x \in y \leftrightarrow x \in z) \to y = z.$

Reading this in English, it says: "If y and z have precisely the same elements, they are the same set" (x, y and z being variables ranging over objects and/or sets).[20] This captures the aforementioned idea that only members matter to the identity of sets.[21]

A consequence of (EP) is that all the sets to which nothing belongs, that is, all sets with no elements, are the same. If there are no winged horses and no unicorns, it can be conjectured that the set of unicorns and the set of winged horses exist; they both lack members. (EP) says that x and y are the same set iff no member of the former is a member of the latter and vice versa – and this is always the case when they are empty. We can therefore talk of *the* empty set, which is usually labeled with the symbol "$\varnothing$".

---

[19]   The formulation provided below is not Frege's, who used what we nowadays would call a higher-order language. This is a complication we can skip here, but, as announced, I shall have something to say on higher-order languages and theories in the following.

[20]   One of the features that render set theory important for mathematics is the fact that sets can be members or elements of sets in their turn. In this sense, sets are not just collections of objects, but objects themselves – taking "object" to mean "something which is capable of set membership." As we will see quite soon, Frege and Russell based their logicist approach on the possibility of reducing numbers to sets by considering them as sets of sets. On the other hand, in many contemporary theories of sets (often called "pure" theories), *only* sets figure in the domain of the theory; there are no *Urelemente*, objects that aren't sets. To put it otherwise, within the range of the things the interpreted theory talks about there are no such things as Gottlob Frege, Juliette Binoche, or the city of Melbourne, and then sets, but only sets. In this case, the only objects the variables of the theory, such as x, y, and z, range over, are sets. These specifications, in any case, are not essential in order to understand what follows.

[21]   Consequently, different properties can originate the same set. To recycle the classical Fregean example: the property of being an animal with a heart and the property of being an animal with kidneys are intuitively different (*having a heart* does not seem to be the same thing as *having kidneys*). However, they correspond to a unique set, for any animal with a heart also has kidneys, and *vice versa*. This is the mark of the difference between such "intensional" entities as properties, and sets.

8   The Abstraction Principle

The second fundamental principle of (what was later to be called) naïve set theory was the *Abstraction* or *Comprehension Principle* (AP). It can be expressed, via the notation with set abstracts introduced above, thus:

(AP1) $x \in \{y \mid P(y)\} \leftrightarrow P(x)$.

Translating into English, this says: "x is a member of the set of the Ps iff x is (a) P" (e.g., Jeffery Deaver is a member of the set of writers if and only if Jeffery Deaver is a writer). Otherwise, it can be formulated without set abstracts, thus:

(AP2) $\exists y \forall x (x \in y \leftrightarrow \alpha[x])$,

where α is a *metavariable* for formulas of the formal language, that is, a placeholder for any of those formulas.[22] Specifically, "$\alpha[x]$" can be replaced by any formula with one free variable x (perhaps occurring more than once), expressing a property or condition on x.[23] (AP2) thus says something like: "There's a (set) y such that, for all x, x is a member of y iff x has the property (satisfies the condition expressed by) $\alpha[x]$." (AP) was supposed to express the idea of set included in Cantor's celebrated definition:

> By an "aggregate," we are to understand any collection into a whole M of definite and separate objects m of our intuition or of our thought.[24]

The basic insight was that *any* "collection into a whole" is a set, which means that any property P in (AP1), or any condition $\alpha[x]$ in (AP2), is taken as defining one. When we think of something as a thing of a certain sort (and taking "sort" in a broad sense, not in the strict sense of

---

[22]   In the following I will use Greek letters, sometimes as (meta-) variables for formulas ("for some formula α ..."), sometimes as names of famous sentences, such as "the Gödel sentence γ ...." Again, context will disambiguate.

[23]   With the proviso: y must not be free in $\alpha(x)$.

[24]   Cantor (1895), p. 85.

sortal concepts), at the same time we appear to think of it as being one of a group, which is itself a thing of a certain sort. More specifically, given any multiplicity with some characterizing condition, the Abstraction Principle seems to guarantee that there exists a set of all and only those objects, and that the set is itself an object. "Object" should be taken as meaning, more or less: something we can refer to as a unity, which is the subject of attributions and predications, and has properties. All this is intuitive if anything is.

But intuition betrays us. In fact, (AP) originated the major crisis in the foundations of mathematics – and the logical milieu in which Gödel grew up. Before we turn to this, we need a few more set-theoretic notions which will prove useful in the following.

9 Bytes of set theory

Another basic relation in set theory is that of *inclusion*, usually expressed by the symbol "⊆". A set, say A, is said to be included in a set B if and only if each element of A is also an element of B. In this case, A is also called a subset of B. For instance: the set constituted by (all and only) the Germans is a subset of the set of Europeans, since all Germans are Europeans. The set of (all and only) AC Milan midfielders is a subset of the set of AC Milan players, and so on.

Never confuse ∈ and ⊆, that is, membership and inclusion. Juliette Binoche is a member of the set of Frenchmen, and since all Frenchmen are European, the set of Frenchmen is included in the set of Europeans. But the set of Frenchmen does not *belong to* the set of Europeans, for it is a set, not a European (the set of Frenchmen, being a set, is an abstract object: it is not a person, therefore it cannot be a European, even though its members, the Frenchmen, are Europeans). The membership relation can hold between objects and sets; the inclusion relation can hold between sets. This does not rule out sets having other sets as members, as I have already said above. The fact that one can have sets whose members are sets, that is, sets of sets, made set theory mathematically significant from Frege and Russell's viewpoint; their basic insight, as we shall see soon, was to define natural numbers precisely as sets of sets.

Another notion we will need in the following is that of *ordered n-tuple*: an ordered couple (or pair), triple, etc. I will write "$<x_1, \ldots, x_n>$"

to signify an ordered n-tuple of objects $x_1, \ldots, x_n$. n-tuples are said to be ordered because, unlike sets, the order in which the elements are listed matters: if x is different from y, $<x, y>$ isn't the same thing as $<y, x>$, whereas $\{x, y\}$ is the same thing as $\{y, x\}$.[25] For instance, the triple <Gottlob Frege, Juliette Binoche, Melbourne> is different from the triple <Juliette Binoche, Melbourne, Gottlob Frege>.

One can define the *Cartesian product* of sets via the notion of ordered n-tuple; let us label it with the spot "·". Given two sets A and B, their Cartesian product A · B is the set constituted by all and only the ordered couples whose first element is a member of A and whose second element is a member of B. Generalizing, given n sets $A_1, \ldots, A_n$, their Cartesian product $A_1 \cdot \ldots \cdot A_n$ is the set of all the n-tuples $<x_1, \ldots, x_n>$ such that $x_1 \in A_1, x_2 \in A_2, \ldots$, and so on. One can build the Cartesian product of a set with itself, and in this case one talks of a *Cartesian power*: given a set A, its Cartesian square, A^2, is the set of all the ordered couples of elements of A. Generalizing, the n-ary Cartesian power of a given set A, A^n, is the set of all the n-tuples of elements of A.

Never confuse Cartesian powers with *power sets*. The power set of a given set A, usually written as "P(A)," is the set of all subsets of A. For instance, given the set {Gottlob Frege, Juliette Binoche}, its power set is the following:

{∅, {Gottlob Frege}, {Juliette Binoche}, {Gottlob Frege, Juliette Binoche}}.

This is because (a) the empty set is included, by definition, in any set, and (b) each set is a subset of itself (although not, as is sometimes specified, a *proper* subset).

10   Properties, relations, functions, that is, sets again

In the following, I shall talk quite often of properties, relations, and functions – mainly, although not only, of properties of numbers, relations

---

[25] The notion of ordered n-tuple can itself be defined in terms of sets, by means of a procedure due to K. Kuratowski. One can define $<x, y>$ as the set $\{\{x\}, \{x, y\}\}$, and $<y, x>$ as the set $\{\{y\}, \{x, y\}\}$.

between numbers, and numerical functions. In mathematics, a function is just a correspondence between one or more numbers called *arguments* (the input), and a unique number called the *value* (the output) given by the function for those arguments (the set of arguments of a function is usually called its *domain*, and the set of values associated with those arguments is usually called its *range* or its *image*).[26] Sometimes people use "function" as a synonym for "operation," and I will follow this usage. For instance, such an operation of elementary arithmetic as addition is a function: when I add 2 to 3 I have a unique value, that is their sum $2 + 3$, that is 5, which corresponds to the two numbers 2 and 3 taken as arguments.

We can now begin to put to work our bits and bytes of set theory: it turns out, in fact, that the notions of (numerical) property, relation, and function can be captured set-theoretically. For instance, the property of being a mammal, or that of being an odd number, can be associated with, or considered as, the sets of objects which enjoy the properties (the set of mammals, the set of odd numbers) and which constitute their extension. Therefore in the following I will often say, indifferently, that an object (an animal, a number) has a certain property, or that an object belongs to the set of all and only the objects having the property (the set of mammals, the set of odd numbers).

Similarly, an *n*-ary relation can be expressed set-theoretically as a set of ordered *n*-tuples: those between which the relation holds. For instance, one can take the binary or two-place relation between father and son (the relation … *is the father of* …) as the set constituted by all and only the ordered pairs whose first member is the father of the second. The binary relation … *is greater than* …, holding between numbers, can be identified with the (infinite) set of all and only the ordered pairs of numbers, $<m, n>$, such that m is greater than n. The ternary or three-place relation … *is halfway between … and …* can be considered as the set of ordered triples such that the first element of the triple is halfway between the second and the third (<Milan, the north pole, the equator>, for instance); and so on.

Also functions can be characterized by means of set-theoretical notions: there exists an easy inter-definability between functions and

---

[26] I'll most often talk of *total* functions, that is, functions that are defined for all arguments, that is, functions that assign a value to all of them. Functions that are undefined for some arguments are called *partial* functions.

sets. With each *n*-ary relation – say R (which, as I have just said, can be considered set-theoretically as a set of ordered *n*-tuples) – one can associate a function with *n* arguments – say c_R – called its *characteristic function*. This is a function, whose range is $\{1, 0\}$, such that:

If $< x_1, \ldots, x_n > \in R$, then : $\quad c_R (x_1, \ldots, x_n) = 1$
Otherwise : $\qquad\qquad\qquad c_R (x_1, \ldots, x_n) = 0 .$

That is to say: the characteristic function c_R for the relation R is the function such that, if R holds between the objects $x_1, \ldots, x_n$ (in this order), then c_R gives 1 as its value for the arguments $x_1, \ldots, x_n$ (in this order); if, on the other hand, the relation does not hold between those objects, then c_R gives 0 as its value. Properties are just a special case: given a set M, corresponding to a property, its characteristic function is the unary function c_M, such that:

If $x \in M$, then : $\quad c_M (x) = 1$
Otherwise : $\qquad\quad c_M (x) = 0.$

That is, it is the function which maps the given argument x to 1 if x has the property at issue (that is, if x belongs to M), and to 0 if x does not have the property at issue (does not belong to M).

Conversely, with each *n*-ary function f one can associate an *n+1*-ary relation G_f, usually called its *graph* relation. This is the relation holding exactly between the arguments of the function and its values, that is, a relation such that:

$< x_1, \ldots, x_n, x_{n+1} > \in G_f$ iff $f(x_1, \ldots, x_n) = x_{n+1} .$

When all's said and done, discourses on properties and relations on the one hand, and functions on the other, are reducible to each other. We can talk only in terms of sets by explaining away functions in terms of their graph relations; or, conversely, we can talk only in terms of functions by explaining away sets (that is, properties and relations) in terms of their characteristic functions.

If you are beginning to wonder what is the purpose of all this apparatus, remember: we're still just setting up the instruments to perform the Gödelian symphony!

11 Calculating, computing, enumerating, that is, the notion of algorithm

Talking about setting up instruments, we shall now take on board another group of definitions, whose importance for our Gödelian piece of music cannot be overestimated. All the notions we will meet in this section are related to that of *algorithm*. Roughly, this is the name given in mathematics to a mechanical (also called *effective*) procedure which, when applied to a number or to a sequence of numbers, terminates after a finite number of steps, providing some information on the number or sequence of numbers.[27] Such a procedure has to be specifiable as a finite series of totally explicit, simple, and deterministic instructions. The instructions must tell you what is to be done at each step of the procedure, so that no creativity, ingenuity, or free choice is required. The procedure is labeled "mechanical" to evoke the idea that a machine, such as a computer, could carry it out: in fact, the connection is so close that I will often use computers as intuitive examples when explaining algorithmic notions.

Now a (say unary) function *f* is said to be (*effectively*) *computable* when there is some algorithm that in principle allows to calculate its value for each of its arguments – that is, when it is possible to specify a series of instructions (for instance, in the form of some computer software) following which one can, in principle, determine mechanically and effectively the output *f(x)* for each input *x*. Generalizing to functions with *n* arguments is straightforward. For instance, addition is a two-argument function which is computable in this sense: at school we learn algorithms, that is, mechanical procedures, to calculate, given two numbers *m* and *n*, their sum *m* + *n*.

A set M is called *decidable* (sometimes also *computable*) if, for every *x*, some algorithm provides a positive or negative answer to the question "*x* ∈ M?" – that is, the algorithm allows one to decide, in principle, if *x* belongs to M or not (for instance, one can set up a computer so

---

[27] Sometimes the term "algorithm" is also used in the literature to name procedures that do not necessarily terminate. As we shall see, discourses concerning algorithms can be applied also to domains that are not, so to speak, "immediately" numerical – thanks to Gödel's work. We will learn these things little by little, in any case.

that, once asked "$x \in$ M?", after a finite amount of time it will answer
with a "Yes" or a "No").

Equivalently, one will claim that a property is decidable in the case
when some algorithm can establish whether a given object x has that
property or not. For instance, we have arithmetical algorithms to decide
whether a given natural number has the property of being divisible by
2, whether it has the property of being prime (i.e., it belongs to the set
of prime numbers, the numbers – bigger than 1 – divisible only by
themselves and by 1), and so on. Also in this case, generalizing to relations is straightforward: an n-ary relation R is decidable if and only if,
for each n-tuple $<x_1, ..., x_n>$, there is some algorithm for deciding ...;
and so on. The terminology is extended to predicates, that is, to those
linguistic entities that denote decidable sets (properties, relations).

One could describe the notions of computable function and decidable set (property, relation) in more verbose and exhaustive ways. But
when the chips are down, it remains the case that we are dealing with
intuitive and, in this sense, slightly vague notions. One may wonder, in
fact, which operations and instructions are intrinsically simple, and
which combinations of such operations or instructions are admissible in
order to preserve the mechanical nature of the procedure. Later in the
development of this book, we will see that the mathematical market
offers different theories taken as delivering precise and systematic
accounts of the notion of algorithm (computable function, decidable
set). We will mainly focus on one of them: the theory of *recursive
functions*. For the time being, however, we can stick to our quick, intuitive characterization. This should be enough to immediately understand
that *a set* (a property, a relation) *is decidable if and only if its characteristic function is effectively computable*. If there exists an algorithm
to decide, for any x, whether x belongs to the given set M or not, then
that algorithm allows us to figure whether the value of the corresponding characteristic function $c_M(x)$ is 1 or 0. Conversely, a function is (effectively) computable if and only if its graph relation is decidable. Discourses
concerning the computability of functions and the decidability of sets
(properties, relations), therefore, can be phrased only in terms of functions, or only in terms of sets (so I will sometimes talk only of sets, sometimes only of functions, and sometimes of both kinds of things).

A set M is called *enumerable* or *effectively enumerable* (sometimes
computably enumerable, or also *semi-decidable*) when there is a
mechanical procedure such that, given some object x, *if* x belongs to M,
the procedure will deliver a "Yes" as its output after a finite amount of

time; but if *x* does *not* belong to M, an answer may not be forthcoming (extending to relations is easy also in this case). This means that, with such sets as M, there is a mechanical procedure that generates all the elements of the set (the objects having the relevant property, etc.). For instance, one can program a computer to compute, and print out one after the other, the (names of the) members of the set – not taking into account the limitations due to time, resources, available memory, etc.

The notions of decidable and of (computably) enumerable or semi-decidable sets are not coextensive: as we shall see, there exist (computably) enumerable or semi-decidable sets that are not decidable (and some such set will play a major role in the development of the Gödelian symphony). The two notions are, nevertheless, closely connected in two important ways.

(1) The first connection is the following: *every decidable set is enumerable*. If we have an algorithm, that is, an effective procedure, to decide in a finite number of steps whether a given *x* is a member or element of a given set M or not, then certainly we also have a mechanical procedure to generate all the (names of) the elements of M one after the other (once again, the generalization to relations is obvious). We can make the point quite simply by resorting, again, to the example from computers: if we can program a computer so that, for each *x*, it can decide in a finite amount of time whether $x \in M$ or not, then we can certainly program it in such a way that (a) the computer prints (the name of) *x* when, once the computing process is done, it turns out that *x* actually belongs to M, and (b) the computer discards *x* if it turns out that *x* doesn't belong to M. This way, the computer will print in sequence, and therefore enumerate in a list, all the elements of M.

(2) To understand the second link between decidable and (computably) enumerable sets, let us consider that, given any set M, by the (set-theoretic) *complement* of M is meant the set of all and only the *x* such that $x \notin M$ (let us label such a complement set "–M").[28] Now

---

[28] Mainstream set theories usually make or presuppose a distinction between the absolute complement of M and its complement relative to a pre-specified set (say U) of which M is a subset. The relative complement of M with respect to U is the set of all elements of U that are not elements of M. The distinction between absolute and relative complement is required because in most of the current theories an absolute complement of a given set M, that is, the set of just anything not belonging to M, cannot be admitted. This is one of the consequences of the set-theoretic paradoxes we are about to meet.

the second connection goes like this: *a set is decidable if and only if both the set and its complement are enumerable.* First, if a given set M is decidable, then its complement −M obviously is too; therefore both sets are computably enumerable because of connection 1 (that is, any decidable set is enumerable). Second, if both a set M and its complement are enumerable, this means that (a) there exists a mechanical procedure to produce all the elements of M in succession - a procedure which sooner or later will let us know (with a "Yes," or by printing its name, etc.) if a given x belongs to M, although it will remain silent if x does not; and (b) there exists a mechanical procedure to produce all the elements of −M in succession - a procedure which sooner or later will let us know if a given x does not belong to M, although it will say nothing if x belongs to it. Therefore a computer will always be able to decide, given some object x, whether $x \notin$ M or not, by combining the procedures (a) and (b), perhaps by alternating one step of the former with one step of the latter. The computer, that is, applies a step of the procedure which enumerates the element of M: if x shows up, x belongs to M and we are home already. Otherwise, it applies a step of the procedure which enumerates the elements of −M, and if x shows up now, then $x \notin$ M. If x didn't show up either way, the computer applies a second step of the first procedure; then a second step of the second one; and so on. Eventually, x will show up as belonging to M or to −M, and the computer will let us know, or print, its positive or negative answer.

I have used several times the expression "in principle." When the general notions of algorithm, computable function, etc., are specified, one typically disregards the factual and practical limitations of the calculus. Pragmatic considerations concerning the time required to perform the calculus successfully, the energy expended in the process, the amount of memory a computer would need to carry it out, or the paper needed to print the output, and so on, are not taken into account: the general, abstract notion of algorithm or effective procedure is, in this sense, deliberately "idealized." Even the systems of symbols actually adopted in the calculus are irrelevant - which does not mean, of course, that calculating with certain notations may never turn out in practice to be more convenient than with certain others. The symbols which designate numbers are called *numerals*, and should not be confused with the numbers themselves. A numeral is not a number, but a linguistic sign designating a number – and the same number, of course, can be designated by different linguistic signs. For instance, in our ordinary written language we can designate the number four with such numerals as "4" or "four." But we could adopt different notational conventions: for instance, we could express the number n by using n strokes, so the number four as | | | | (this is the so-called *tally* notation). Now, it is certainly easier for us to add 34 to 8 than XXXIV to VIII. But translating from one notational system to another is itself an effective procedure, so the fact that numbers (arguments, values, etc.) are presented in some numeric notation or other changes nothing from the viewpoint of the idealized notion of algorithm.

12   Taking numbers as sets of sets

Back to history now. We have seen how Peano axiomatized arithmetic via the three basic notions and the five axioms which now bear his name. You should remember that the notion of (natural) number was taken as a primitive, intuitive one. But the logicist enterprise pursued by Russell and Frege required numbers themselves to be reduced to something even more fundamental and basic: specifically, it required numbers to be definable in terms of sets, and of properties of or relations between sets. The basic idea was to the effect that numbers be taken as properties of sets, and therefore (given that, as we have seen, properties can be reduced to sets in their turn) as sets of sets.

To grasp the basic intuition, we have to buy the following definition: two given sets A and B are said to have the *same cardinality*, or to be *equinumerous*, when it is possible to put their elements in a one-to-one connection – that is, it's possible to pair each element of A with one and only one element of B, so that nothing is left "unpaired." In this case, one also claims that there is a one-to-one correspondence or mapping between the two sets, or a *bijective* function, or bijection, pairing each element of A to each element of B. For instance, assume that A is the set of all husbands, and B is the set of all wives, and assume also that the only admissible marriage prescribes monogamy. Then we know that A and B have the same number of elements, even when we don't know exactly what that number is, that is, how many married couples there are. The reason is precisely that we know we can pair one-to-one each husband to the respective wife, so that no husband remains without wife, and no wife without husband: there is a one-to-one mapping between the two sets.

Now we can define the *number of* a given set as the set of all sets equinumerous to it. Any number is then characterized as a property of sets which have the same cardinality, and therefore as a set of equinumerous sets. For instance: which feature is shared by the set of Hercules' labors, the set of Apostles, and the set of months in a year? *Twelve* immediately comes to mind, and the number twelve can be seen as the property of all and only the sets which are equinumerous to those sample sets, and therefore as a set of sets. When we claim that the months in a year are twelve, we are attributing a property not to the months taken individually (January, February, etc.) but to their set: it's a property of that set, and of various others (such as the set of Apostles, etc.); so it's a set of sets. This is how discourses on numbers can be reduced to discourses on sets of sets. And the notions of set, and of belonging to a set, unlike that of number, appear to be definitely primitive ones, irreducible to anything still more fundamental.

This reduction of numbers to sets was just the first half of the story for the logicist project. The second half was to consist in obtaining all of mathematics from the primal set-theoretic notions, taken in those times as logical concepts, by means of purely logical inferences. Logicists aimed at deriving mathematics from logic via deductive chains going from the premises of symbolic logic, down to geometry, through finite and infinite arithmetic.

But the logicists' confidence was doomed to be deeply shaken.

13   It's raining paradoxes

Russell discovered that set theory, which was supposed to provide the foundational machinery for arithmetic (and therefore for the whole of mathematics), actually provided a devastating contradiction. In fact, set theory quickly found itself caught in a storm of contradictions. Let us see how.

All the things I have been saying so far on sets will be needed in the remainder of this book; and almost all the set-theoretic notions presented so far are nowadays customary mathematics. But at least one of the principles we have met brings trouble. This is the Abstraction or Comprehension Principle (AP): no matter exactly how one phrases it, (AP) appears to grant that to any property or condition there corresponds a set. This seemingly intuitive and obvious idea produces various set-theoretic paradoxes in the so-called naïve (version of) set theory. These paradoxes struck the foundations of mathematics as an earthquake at the beginning of the twentieth century.

The simplest and most celebrated of the set-theoretic paradoxes hit Frege on June 16, 1902, in the form of a letter sent by Russell – a letter which belongs to the history of contemporary thought. The paradox exploits a bunch of beautifully simple insights. First, it is intuitive that some sets do not belong to themselves – do not include themselves as members or elements. The set of Frenchmen, for instance, is not French itself (it's not a person of French nationality, for it is a set, an abstract object). Therefore such a set does not belong to the set of Frenchmen. The set whose only member is Juliette Binoche is not Juliette Binoche (a French actress, thank God, is not a set), so it does not belong to itself either. But, still intuitively speaking, it seems that some sets *do* belong to themselves: the set of all sets with more than one element, for instance, has more than one element, so it should be a member of itself. The set of anything but Juliette Binoche is not Juliette Binoche (thank God), so it should also be a member of itself. (Don't you scent in these self-membered sets the fragrance of self-reference? Recalling the Liar, you should also smell the danger.)

The sets which don't belong to themselves are often called "normal." This leads us naturally to consider the set of all normal sets, which is usually called "R" after Russell:

$$R = \{x \mid x \notin x\}.$$

Translating into English: "R is the set of all and only those things x that are not members of themselves." This set of non-self-members originates the paradox that caused Frege's sorrow – with the decisive help of the Abstraction Principle. Since the schematic $\alpha[x]$ in (AP2) stands for any condition or property, we can take precisely the property of *not being a member of oneself*, $x \notin x$, and we get:

$$\exists y \forall x (x \in y \leftrightarrow x \notin x).$$

Translating into English:"There is a (set) y, to which any (set) x belongs if and only if x does not belong to itself." So there exists a set and, by the Extensionality Principle, *the* set, corresponding to such a condition, i.e., y is precisely R:

$$\forall x(x \in R \leftrightarrow x \notin x).$$

Translation:"For all x, x belongs to R if and only if x does not belong to itself." Now, R in its turn is something about which we can speculate, given any property or condition, whether it has that property or satisfies that condition, or not. This is the case also with the property of not being a member of oneself. Since what holds for any x holds for R, we have:

$$R \in R \leftrightarrow R \notin R,$$

that is, R belongs to itself if and only if it doesn't. This is "*the* contradiction," as Russell called it (it has the shape of a biconditional, but we easily get an explicit contradiction of the form $R \in R \wedge R \notin R$ via elementary logical steps). It follows via simple reasoning from the Abstraction Principle, which was assumed to be a quite obvious basic principle of set theory. Logicists discovered that if (AP) is assumed without restrictions, allowing any property or characterizing condition to deliver the corresponding set, then the consideration of some properties, such as non-self-membership, leads straightforwardly to paradoxes.

14   Cantor's diagonal argument

Russell's paradox is a simplified variant of a paradox deducible within naïve set theory and known to Cantor since 1899, even though it was published only in 1932. This begins with consideration of the *universal* set, which is most often indicated (after Peano) as V. V is usually characterized by means of a condition anything is expected to satisfy, such as self-identity:

$$V = \{x \mid x = x\}.[29]$$

[29]   See e.g. Fraenkel, Bar-Hillel, and Levy (1973), p. 124.

This is just the set of everything. But V can be taken as the set of all *sets* if we take a pure theory of sets, that is to say, if we assume that the domain described by the theory does not include *Urelemente*, objects that are not sets. It is not difficult to have V afford us a contradiction, via a line of reasoning just slightly more complex than that involved in Russell's paradox. To obtain this new paradox, one has to consider *Cantor's Theorem*: the fundamental theorem due to Cantor, claiming that the power set P(A) of any given set A (that is, as we know, the set of all subsets of A) has larger cardinality than (so is "bigger than") A: P(A) > A.

The key to the theorem lies in Cantor's ingenious construction called the *diagonal argument*. Cantor initially used the argument to show that the set of natural numbers is not the largest infinite set, for it is exceeded by the set of real numbers (informally, the numbers represented by an infinite decimal expansion, such as the famous π: 3.141593...). Before Cantor, mathematicians were aware of the fact that the real numbers are somehow more numerous than the naturals,[30] even though there are infinitely many natural numbers. But to make full sense of the intuition one has to clarify the idea that one infinity can be "greater" than another, and therefore not "equinumerous" with it. It was precisely the idea of equinumerous sets, that is, of sets whose elements can be paired one-to-one via a bijective correspondence, that provided the required clarification.

Cantor's diagonal argument begins by assuming, for the sake of a *reductio ad absurdum* (that is, a refutation of the assumed thesis), that there is such a one-to-one mapping between natural and real numbers. An infinite set whose members can be paired one-to-one with the naturals is said to be *countably infinite*, or *(d)enumerably* infinite, or denumerable. The elements of such a set can theoretically be arranged in a list – an infinite one, of course, and therefore one that we could never finish writing down in practice, but such that (the name of) every member of the set will appear sooner or later in the list, an acceptable list being such that each member appears as the *n*th entry for some finite *n*.

Now, let us assume that the set of real numbers is enumerable. This means that we could have a list of all the (numerals for) real numbers,

[30] Or than the rational numbers, which, despite having the property of *density* (that is, the property that between any two rationals there sits another – infinitely many others, in fact) are "as many as" the naturals.

which would look like an "infinite square" or matrix, such as the following:

1/3	=	0.	**3**	3	3	3	...
1/2	=	0.	5	**0**	0	0	...
√0.1	=	0.	3	1	**6**	2	...
√0.5	=	0.	7	0	7	**1**	...
...	=	...	...	...	...	...	...

Of course, we could list the real numbers in a different order, thereby having different squares, but this is irrelevant. What matters is that, were such a list possible, we would have enumerated them all. Then we could easily have each item correspond one-to-one to a natural number. But consider the real number – call it r – whose decimal expansion is as follows: the first decimal digit is equal to the decimal digit of the first number in the list, increased by 1 (when we have a 9, we always turn it into a 0); the second digit equals the second digit of the second number in the list, increased by 1; ... ; the nth digit equals the nth digit of the nth number in the list, increased by 1; and so on. We are interested in the bold-faced decimals in the square (which make us see why Cantor's argument is labeled "diagonal"). So the number at issue is $r = 0.4172...$. Now r cannot be in the list: for it differs from the first number in the list at least in the first decimal; it differs from the second at least for the second decimal; ...; from the nth at least for the nth decimal; and so on. The list does not include r, and so is incomplete, against the initial assumption. The procedure holds for whichever way one tries to constrict the real numbers in a list: we can always produce an element (whose identity, certainly, will vary according to the way in which the list is constructed) that cannot appear as an item in the list. All in all, the set of real numbers is not enumerable: it is actually larger than the set of natural numbers.

But this is just the beginning. Given *any* infinite set, Cantor's Theorem in its general version tells us that we can always have a larger infinity: we just have to consider the power set of the infinite set we began with.[31]

---

[31] Here is a general account. I have said that two sets x and y have the same cardinality (let's write: $x \cong y$), or are equinumerous, iff there is a one-to-one correspondence between them: a function mapping each member of x to a member of y, and such that

Now consider the universal set V, i.e., the set of all (pure) sets, and take its power set P(V). Since all members of P(V) are sets, P(V) is a subset of V. But V is itself a subset of P(V). Therefore, P(V) = V. So there is a one-to-one correspondence between V and P(V) (namely, identity), and P(V) $\cong$ V. But Cantor's theorem rules this out for any set, so we have:

$$P(V) \cong V \land \neg(P(V) \cong V).$$

Even more rapidly: given Cantor's theorem, P(V) is bigger than V. This is inconsistent with the fact that V is, by definition, the most

---

each member of y is mapped to by a single member of x. Then, x is *bigger than or equal to y* ($x \geq y$) iff there is a subset of x which has the same cardinality of y; and x is bigger than y ($x > y$) if $x \geq y$ but it is not the case that $x \cong y$. Now Cantor's Theorem says that there are no functions from x *onto* its power set P(x) (the set of all subsets of x), that is to say, P(x) has a greater cardinality than x: P(x) > x. It is easy to see that P(x) $\geq x$; the tricky part consists in showing that it is not the case that P(x) $\cong x$. Cantor's proof begins by assuming – again, for the sake of a *reductio* – that there is a one-to-one function ϕ from x to P(x), so that they have the same cardinality. Now the diagonal argument goes as follows: consider the set z of all the elements of x that are not members of the set assigned to them by ϕ – so $z = \{y \in x \mid y \notin \phi(y)\}$. z is a member of P(x), since it is a subset of x. So there must (by supposition) be an element w of x, such that $z = \phi(w)$. The question is: is w a member of z, i.e., $\phi(w)$, or not? We have:

$w \in \phi(w) \leftrightarrow w \in \{y \in x \mid y \notin \phi(y)\} \leftrightarrow w \notin \phi(w)$.

Given the Law of Excluded Middle, either w is a member of $\phi(w)$ or not, hence:

$w \in \phi(w) \land w \notin \phi(w)$.

The contradiction so deduced leads us to conclude that there cannot be such a one-to-one mapping. Roughly, the strategy is always the same: given a group of objects of a certain kind, the diagonal construction allows us to define an object that cannot be in the group "by systematically destroying the possibility of its identity with each object on the list. The new object may be said to 'diagonalise out' of the list" (Priest (1995), p. 119), as happened with our allegedly complete enumeration of the real numbers with respect to the number r, which was so constructed as to be distinct from each item in the list.

inclusive of all sets: V would have to be bigger than itself! And this is Cantor's paradox.[32]

Another paradox I shall introduce very quickly is Burali-Forti's. This is important both historically, since it was the first to be discovered, and theoretically, since its proof is a direct one (no excluded middle is required). The paradox concerns ordinal numbers. Cantor's initial idea was that ordinals should index *well-ordered* sets. A well-ordered set is a set such that each of its non-empty subsets has a least element (following von Neumann's later idea, an ordinal can correspond to the set of the preceding ordinals: so if 0 is $\varnothing$, 1 is $\{0\}$; 2 is $\{0, 1\}$; 3 is $\{0, 1, 2\}$; and so on). Now consider the set Ω of *all* ordinals. One can give independent arguments for both $\Omega \in \Omega$ and $\Omega \notin \Omega$. By construction, Ω is itself well-ordered, so since any well-ordered set has an ordinal number, Ω must have an ordinal too. However, this ordinal must be greater than any member of the set, and therefore it cannot be in the set.[33]

15   Self-reference and paradoxes

I have dealt with the details of the proofs quite quickly and informally, but this rapid presentation of the most famous set-theoretic paradoxes might nevertheless look a bit technical. What matters to us is that these paradoxes were at the origin of the crisis in the foundations of mathematics at the beginning of the twentieth century. The correspondence with Dedekind shows that Cantor was aware of the paradoxes – particularly of the fact that the universal set was an anomaly with respect to the diagonal argument. I should also add that he wasn't too bothered. Being a religious man, he had a certain tendency to see the whole situation with a mystical eye:

---

[32]   Thus Fraenkel, Bar-Hillel, and Levy (1973), p. 7. Cantor's paradox is nothing but Russell's, once one chooses as ϕ the identity function (therefore, sometimes scholars speak of a unique Cantor–Russell paradox). In this case, $\phi(w)$ just *is* w, i.e., $\{y \in x \mid y \notin y\}$, and the thing goes like:

$$w \in w \leftrightarrow w \in \{y \in x \mid y \notin y\} \leftrightarrow w \notin w.$$

[33]   See ibid., p. 8.

I have no doubt at all that in this way we extend ever further, never reaching an insuperable barrier, but also never reaching any even approximate comprehension of the Absolute. The Absolute can only be recognized, never known, not even approximately.[34]

But Russell (who was not a man of faith) saw the implications of Cantor's theory of infinite sets and numbers going down in flames. I have claimed that the distinction between semantic and set-theoretic paradoxes, due to Ramsey, came after the publication of Russell and Whitehead's *Principia mathematica*. Russell believed that all the logical paradoxes had their root in some form of circularity, or self-referentiality, which he had named "reflexiveness":

> In all the above contradictions ... there is a common characteristic, which we may describe as self-reference or reflexiveness. The remark of Epimenides must include itself in its own scope. If *all* classes, provided they are not members of themselves, are members of [R], this must also apply to [R]; and similarly for the analogous relational contradiction ... In the case of Burali-Forti's paradox, the series whose ordinal number causes the difficulty is the series of all ordinal numbers. In each contradiction something is said about *all* cases of some kind, and from what is said a new case seems to be generated, which both is and is not of the same kind as the cases of which *all* were concerned in what was said.[35]

Russell's solution came with his *theory of logical types*, which he proposed in various papers, and incorporated in the *Principia*. The theory developed a rigid hierarchy of types of objects: individuals, sets, sets of sets, sets of sets of sets ... (something quite similar to the Tarskian hierarchy of metalanguages).[36] What belongs to a certain logical type can be (or not be) a member only of what belongs to the immediately superior logical type. The membership relation can hold, or fail to hold, only between an individual and a set of individuals; or between a set of individuals and a set of sets of individuals; and so on. The construction allows any set to contain only things of one order: it allows only sets composed, so to speak, of objects that are homogeneous with respect to the hierarchy. Therefore, there is no set of all sets, or of all ordinals, etc. Such sets

[34] Hallett (1984), p. 42.
[35] Russell and Whitehead (1910-25), pp. 61-2.
[36] Of course, the historical succession is reversed with respect to my exposition. Tarski presented his theory of truth and his hierarchic approach several years after Russell.

would have to be constituted by members of totally heterogeneous kinds, that is, by things belonging to different levels in the hierarchy. As is clear, the whole construction aims at ruling out self-referential expressions such as "$x \in x$", or "$x \notin x$". These are now rejected as ill-formed: they are taken as simply meaningless. For instance, Russell's paradox disappears because a set can neither be, nor not be, a member of itself. For the same reason, Cantor's paradox disappears because there can be no set V of all sets: we cannot even *say* within the theory that a set contains all sets.

Other rigorously axiomatized theories of sets, developed at the same time as Russell's as well as later, are based upon a general principle that has come to be called the principle of the *limitation of size*, and which, roughly, prohibits some very comprehensive sets. The mainstream and most popular axiomatic set theory nowadays is that proposed by Zermelo, developed and modified by Fraenkel, known as **ZF**, or as **ZFC**,[37] depending on whether or not it includes a set-theoretic axiom called the *Axiom of Choice* (but we can skip the details). Other axiomatic theories, such as those proposed by von Neumann, Bernays, and Gödel himself, introduce a distinction between *sets* and *classes*.[38] It is claimed that classes, as the extensions of some very comprehensive predicates, cannot be taken as full-fledged mathematical objects capable of set membership. Things can be members of sets or classes, and sets can be members of classes, but classes can be members of nothing. Most of these theories, despite avoiding the known paradoxes, have to abandon intuitively plausible or philosophically important sets, such as the total set V (an exception is Quine's system **NF**, which however is not particularly popular among mathematicians). In all the mainstream accounts, the Abstraction-Comprehension Principle has to go, and its work is now carried out as far as possible by weaker and more limited principles. But we don't need to go into the details of axiomatic set theories, for now we have more or less all the set-theoretical notions required to understand the rest of this book.

Before we begin to play the Gödelian symphony, however, we have to take into account another chapter in the story of the crisis in the foundations of mathematics, concerning a development indispensable to an understanding of Gödel's position: the advent of Hilbert's Program.

---

[37] In the following I will use boldfaced capital letters to label famous formal systems and theories.

[38] The terminology is not uniform: sometimes in this context sets are also called *classes*, and classes in the strict sense are called *proper classes*.

2

Hilbert

At the Second International Congress of Mathematicians held in Paris in 1900, David Hilbert produced one of the most famous talks in the history of mathematics. Hilbert was the greatest mathematician of his time. Just one year before, he had published the *Foundations of Geometry*, which was considered the most important work on the subject since Euclid. His talk included a list of 10 "open problems" of mathematics, and this was subsequently expanded to 23. The list was so important that, since then, brilliant mathematicians aiming at the Fields medal – that is, the equivalent for mathematics of the Oscar – have often addressed one of (the still unsolved among) Hilbert's Twenty-Three Problems.

The Second Problem in the list was: prove that arithmetic is consistent.

1   Strings of symbols

After the events described in the previous chapter, the issue of the foundations of mathematics had become more and more pressing for this champion of mathematics:

> Let us admit that the situation in which we presently find ourselves with respect to the paradoxes is in the long run intolerable. Just think: in mathematics, this paragon of reliability and truth, the very notions and inferences, as everyone learns, teaches and uses them, lead to absurdities. And where else would reliability and truth be found if even mathematical thinking fails?[1]

[1] Hilbert (1925), p. 374.

Hilbert decided to address the crisis in the foundations of mathematics. He did it in numerous writings published between 1900 and 1931, where he proposed a kind of foundation labeled, for reasons we are about to see, *formalism*. He elaborated a program allegedly capable of solving all the problems mathematics had faced because of the emergence of the paradoxes in Cantor's set theory.

Hilbert's Program consisted of two stages. First, Hilbert proposed to produce a complete formalization of arithmetic. He proposed, that is, to translate the logical and arithmetical principles to be used in the development of a significant portion of classical mathematics (which, as we already know, was proved reducible to arithmetic) into a *formal system*.

But what is a formal system? In the first instance one might say that, in Hilbert's view, a formal system resembles a classical axiomatic theory. The decisive difference is that the theory the formal system formalizes has been translated into a rigorous artificial language of symbols and, above all, has been deprived of any reference to such things as meaning and truth. The principles of the system have to be specified and considered merely in terms of the form of the linguistic objects, not taking into account their content or meaning – hence the label "formalism." But let us listen to the key idea of the strategy from Hilbert's own voice:

> We now divest the logical signs of all meaning, just as we did the mathematical ones, and declare that the formulas of the logical calculus do not mean anything in themselves.... In a way that exactly corresponds to the transition from contentual number theory to formal algebra we regard the signs and operation symbols of the logical calculus as detached from their contentual meaning. In this way we now finally obtain, in place of the contentual mathematical science that is communicated by means of ordinary language, an inventory of formulas that are formed from mathematical and logical signs and follow each other according to definite rules. Certain of these formulas correspond to the mathematical axioms, and to contentual inference there correspond the rules according to which the formulas follow each other; hence contentual inference is replaced by manipulation of signs according to rules, and in this way the full transition from a naïve to a formal treatment is now accomplished.[2]

---

[2] Ibid., p. 381.

That is to say: a completely formalized axiomatic system, in Hilbert's view, should be considered merely as a system of signs. We have the primitive *symbols* of the artificial language, specified at the outset; the *formulas*, finite strings of symbols concatenated in accordance with some syntactic or "grammatical" rules, themselves rigorously specified so as to leave no room for doubt on which strings of symbols are grammatically correct or, as logicians say, "well-formed"; the *axioms*, specific (kinds of) formulas whose set is rigorously delimited in its turn; the *rules of inference*, instructions specifying the admissible ways of operating on the well-formed strings in order to obtain new formulas from the initial ones; the *formal proofs*, finite sequences of formulas, each of which has to be either an axiom, or a string obtained from the previous formulas via the rules of inference; and the *theorems* – the formulas which are said to be provable in the system – which are just the final lines in the proofs.

As you should remember, some of these notions (such as those of axiom, proof, and theorem) had already appeared in the traditional axiomatic method. Specifically, such axioms as Euclid's or Peano's had been taken as manifest truths of geometry or arithmetic. However, in the formalist's account of these notions, axioms and formal systems are not considered descriptive of anything. The axioms being deprived of any meaning, the deductions of formulas within the system are taken as a kind of mechanical, purely combinatorial procedure. Deductions are just manipulations of strings of symbols that begin with the axioms and, via the transformations allowed by the rules of inference, produce the theorems. The relations of dependence between the formulas (specifically, between the axioms and the theorems) should thus be fully "visible": their properties and features can be read off from the purely syntactic and structural connections between (the shapes of) the strings.

However, on a formal system considered in this way – say **S** – and on its artificial, formal language – say **L** – whose expressions have been deprived of meaning, one can, of course, make meaningful claims. One can remark that a certain formula (that is, a certain finite string of symbols) is longer than another; or that a certain symbol occurs n times within a formula; or that a formula is a theorem of **S** (which, in the formal context, simply means: it is the last one in a list of strings, each of which is either an axiom of **S**, or a formula obtained from the preceding formulas via manipulations allowed by the rules of inference); or that a certain finite list of strings is a proof in **S** – that is, a finite sequence of formulas; and so on.

Now, such statements as "The formula '$x + 0 = x$' is a theorem of **S**," or "Such and such a sequence of formulas is a proof in **S**," belong to what Hilbert called *metamathematics*. The terminology is easily understood: these statements have no place in the formal system **S** at issue, considered as a formalization of (a certain portion of) intuitive mathematics, but whose language **L** is taken as uninterpreted. They are claims made in (more or less) ordinary English, and in talking about the formal system at issue (this is why they are *meta*mathematical). Specifically, they talk about the *syntax* of the system. They ascribe to certain expressions of **S** purely syntactic features: properties specifiable without referring to meanings and semantic notions, for they concern only the form of the symbols of those expressions. For instance, the aforementioned statement:

(1)   The formula "$x + 0 = x$" is a theorem of **S**

ascribes to a formula of the system a syntactic feature. Being a theorem of **S**, in the formalist approach, is a property a formula has when it is the last one in a sequence of strings, such that ..., and so on. And, of course, such a specification only mentions symbols and their modes of combination, not their interpretations, nor any issue concerning meaning and semantics.

Clearly, the distinction between expressions of a formal system **S** and metamathematical statements on those expressions is quite close to the distinction between a language and a (its) metalanguage, which we encountered in the previous chapter. On the one hand, the sentences of metamathematics don't belong to the formal language **L** of the system: sentence (1) is not a formula of **L**. On the other hand, the expressions of **L** are, typically, mentioned within the metamathematical claims – which is made evident by the quotation marks: in (1), the formula "$x + 0 = x$", which belongs to the formal language, is mentioned, not used.

2   "... in mathematics there is no *ignorabimus*"

Hilbert considered metamathematics as a truly new area of mathematics – as the discipline which deals with what can and cannot be proved in mathematics. Therefore, metamathematics was also called *proof theory* (and a significant portion of the book you are now reading is, in fact, metamathematics or proof theory). The distinction between a system **S** formalizing (a fragment of) mathematics, and its metamathematical description in terms of proof theory, was central to the second stage of Hilbert's Program. He proposed, in fact, to solve the crisis in the foundations of mathematics by *proving metamathematically the consistency of formalized arithmetic via purely finitary methods*. Let us see what this means – particularly, what is meant by "proving the consistency" and by "finitary."

A formal system **S** is said to be (simply) *consistent* if, for any formula α of the formal language **L** it is built on, the system does not allow one to prove both α and its negation, therefore a contradiction. If, on the contrary, this is the case for some α, **S** is said to be inconsistent. In symbols: **S** is consistent iff it is not the case that $\vdash_s \alpha$ and $\vdash_s \neg \alpha$.[3] This is sometimes called (simple) *syntactic* consistency,[4] for it is a notion which, in a formal context, has to be taken as specified in terms of the syntax of the system, with no reference to meanings. To claim that a given system **S** is consistent is just to claim that, given any well-formed finite string of symbols which is a formula α of the system, it cannot happen that both (a) (an occurrence of) the string α, and (b) (another occurrence of) that string with a "$\neg$" prefixed to it, are theorems (that is, as usual: it cannot happen that both formulas are the final lines in two finite sequences of formulas which constitute two formal proofs in **S**). Proving that the formal system formalizing arithmetic is consistent amounts to proving

---

[3] The symbol "$\vdash$" is sometimes called the sign of assertion, and is due to Frege. Its meaning is that what follows it is a theorem *of* a certain formal system, which, if needed, can be specified after the symbol itself. Given a formal system **S**, "$\vdash_s$" indicates that the formula following it is a theorem of **S**, that is, a provable formula of **S**. In contemporary proof theory, it does not make sense to wonder whether a formula is provable *simpliciter*. What makes sense is to wonder whether a formula is provable in a specific formal system (and a formula which is provable in a system may be unprovable in another; this is a point I shall come back to). Finally, it should be noticed that "$\vdash$" (typically) is not a symbol of the formal language of the system: it is a metalinguistic sign, used in order to make a metalinguistic claim on a formula – in order to attribute to it precisely the syntactic property of being a theorem.

[4] In Gödelian contexts, one talks of simple consistency to distinguish it from another kind of (slightly more technical) consistency, called *omega-consistency*. This notion was introduced by Gödel in his proof of the Incompleteness Theorem, and will be explained later.

metamathematically that two such configurations of symbols can never be derived from the axioms of the system by means of its rules of inference.

What then is meant by the claim that a consistency proof for formalized arithmetic has to be given via purely finitary methods? Things become a bit blurry here, because, even though he provided some examples, Hilbert was never totally explicit in defining what was meant by "finitary methods." The German word used by Hilbert was *finit*, which was turned into *finitary* (as far as I know, by Stephen Kleene); another equivalent term often used is *finitistic*. Undoubtedly, Hilbert wanted these methods for proving the consistency of formalized arithmetic to be absolutely *safe*. An initial, intuitive characterization can be provided by saying that such methods should not involve in any way the concept of actual infinity.[5] They should refer not to actually infinite objects but, at most, to potentially infinite collections, such as the sequence of natural numbers 0, 1, 2, …, considered as indefinitely extendible, but not as a completed whole. As we have seen in the previous chapter, the unreliable Cantorian infinite sets were indeed at the heart of the set-theoretic paradoxes which had subverted the logicist project. Furthermore, between the nineteenth century and the twentieth, various unorthodox mathematicians (mainly the *intuitionists*, guided by L.E.J. Brouwer) had rejected (much of) the so-called "infinitary" or infinitistic mathematics by declaring its theorems fallacious or utterly meaningless. The criticisms advanced by the intuitionists dealt precisely with the notion of infinity, and especially with the "higher infinities" whose existence had been allegedly attested by the problematic Cantorian theory. Therefore, Hilbert proposed to dispense with these notions by adopting only finitary concepts and reasoning in the proof of the consistency of formalized arithmetic.

During the twenties, Hilbert developed the "consistency program" into the "conservativity program." The details are not essential to our main path but, roughly, the story goes as follows. Hilbert proposed to consider the so-called "real" mathematical statements, that is, those that do not refer to actual infinities, as autonomously meaningful; to treat, instead, sentences which are not finitary in this sense as "ideal," not provided with self-sufficient content; and to show that infinitary or "ideal" arithmetic is a *conservative extension* of "real" arithmetic – which, in

---

[5] As claimed by Kleene (1976), (1986).

this context, means that anything that can be proved ideally can also be proved really, albeit perhaps in more roundabout ways. Orthodox infinitary mathematics for Hilbert would have turned out to be something like a detour: the shortest route to obtain finitistically provable results. And the metamathematical proof of this fact had to be provided itself by finitary means, for, had the proof employed methods whose mathematical legitimacy was under debate, the result would obviously have left the critics unsatisfied.

Now, the formalist approach was seemingly capable of guaranteeing such a restrictive finitistic rigor in the metamathematical proofs. The crisis in the foundations of mathematics had had its origin in the fact that mathematicians had incautiously climbed the ladder of abstraction and idealization through more and more complex transcendent notions – until they had found themselves in the dead end of the paradox. In metamathematics or proof theory, however, we are not concerned directly with mathematical objects (such as numbers, functions, sets, etc.). Instead, we are dealing with *symbols*: the formulas and axioms of a formal system, in which the mathematical concepts have been formalized. And when we deal with such systems and their artificial languages in purely syntactic terms, we have to work only with finite objects, never with completed infinities: symbols, finite sequences of symbols such as the formulas, and finite sequences of formulas such as the formal proofs. No matter how abstract and transcendent some mathematical concepts may turn out to be, it seems that, by looking only at the shapes of the symbols and of the strings in a fully formalized system, we can keep under control – we can literally keep our eyes on – the syntactic features of the system involved in the metamathematical reasoning on it. By examining the structural properties of the expressions and formulas, we should be able to show that the system is consistent: it is such that we will never obtain in it the proofs of two reciprocally contradictory formulas.

Hilbert's Program allegedly embodied the assumption that a formal system for arithmetic, besides being provably (in the finitary sense) consistent, had to be capable of deciding the mathematical problems expressible in its underlying formal language. This was but the metamathematical counterpart of Hilbert's more general persuasion concerning the solvability, in principle, of *any* mathematical question. Many classical mathematical problems, even when they are easily formulated, can be very complicated to solve (at the outset of this book I mentioned Fermat's Theorem, whose proof recently completed by

Andrew Wiles is the culmination of a complicated collection of mathematical techniques involving notions of algebra and number theory, elliptic curves, modular forms, etc.). However, after claiming that "our proof theory forms the necessary keystone in the edifice of axiomatic theory," our confident formalist mathematician added:

> As an example of the way in which fundamental questions can be treated [within proof theory] I would like to choose the thesis that every mathematical problem can be solved. We are all convinced of that. After all, one of the things that attract us most when we apply ourselves to a mathematical problem is precisely that within us we always hear the call: here is the problem, search for the solution; you can find it by pure thought, for in mathematics there is no *ignorabimus*. Now, to be sure, my proof theory cannot specify a general method for solving every mathematical problem; that does not exist. But the demonstration that the assumption of the solvability of every mathematical problem is consistent falls entirely within the scope of our theory.[6]

3   Gödel on stage

It is at this point of the story that Gödel takes the stage: Gödel who, as a young, gifted mathematician, decided in 1929 to address the list of Hilbert's Problems – and chose the Second; Gödel who, from the start, had the intention not of destroying Hilbert's Program but of moving it forward; and Gödel who, several years later, ascribed his astonishing successes in mathematical logic to his going deliberately against the tide, unmoved by the general disbelief in infinitary reasoning displayed by the mathematicians of his time.

Let us listen to the opening of Gödel's paper – to the beginning of the symphony. Gödel's words are loaded with references. Now that we have tuned our instruments in the previous and current chapters, though, we should be able to appreciate all the suggestions in his memorable *incipit*:

> The development of mathematics toward greater precision has led, as is well known, to the formalization of large tracts of it, so that one can

---

[6]   Hilbert (1925), p. 384.

prove any theorem using nothing but a few mechanical rules. The most comprehensive formal systems that have been set up hitherto are the system of *Principia mathematica* (PM) on the one hand and the Zermelo–Fraenkel axiom system of set theory (further developed by J. von Neumann) on the other. These two systems are so comprehensive that in them all methods of proof today used in mathematics are formalized, that is, reduced to a few axioms and rules of inference. One might therefore conjecture that these axioms and rules of inference are sufficient to decide any mathematical question that can at all be formally expressed in these systems. It will be shown below that this is not the case, that on the contrary there are in the two systems mentioned relatively simple problems in the theory of integers that cannot be decided on the basis of the axioms. This situation is not in any way due to the special nature of the systems that have been set up but holds for a wide class of formal systems; among these, in particular, are all systems that result from the two just mentioned through the addition of a finite number of axioms … The precise analysis of this curious situation leads to surprising results concerning consistency proofs for formal systems.[7]

4   Our first encounter with the Incompleteness Theorem …

The "curious situation" Gödel referred to was precisely that produced by his Theorem. In the first section of the paper, he began by sketching "the main idea of the proof, of course without any claim to complete precision."[8] I shall now re-examine that idea within a framework analogous to the original Gödelian one – "analogous" in the following sense. First, I'm providing an informal explanation of the basic structure of Gödel's proof. Second, the explanation will be based on a *semantic* approach – which, in this context, means: an approach appealing to semantic concepts and, specifically, to the notion of truth. The exposition will constitute our first encounter with Gödel's Theorem, and is similar to those you will find in several introductory textbooks.

However, it should be taken *cum grano salis*. Philosophers have often had in their sights only the initial explanation carried out "without

---

[7]   Godel (1931), pp. 17–19.
[8]   Ibid., p. 19.

any claim to complete precision" in the first section of Gödel's 1931 paper; and this has generated various misinterpretations of the incompleteness results (according to some, Wittgenstein himself was misled in this way in his famous comments on Gödel's First Theorem in the *Remarks on the Foundations of Mathematics*). To avoid misunderstandings, one has to look at the various stages of Gödel's proof in slow motion, that is, with all the due mathematical (and, above all, metamathematical) details. This produces a number of complications; but also, this is what Gödel performed in the rest of his paper, and what will be carried out again in the subsequent chapters (above all, we will learn the complications little by little).

Let us begin by defining four notions that constitute two couples of properties of formal systems – a semantic and a syntactic one: (1) (syntactic) consistency; (2) syntactic completeness; (3) (semantic) soundness or correctness; and (4) semantic completeness. I shall refer to these four notions quite often throughout this book.

(1)   Actually, we already know the first property: a system **S** is said to be (syntactically) consistent when, for any given formula α of its language **L**, **S** does not prove both the formula and its negation $\neg\alpha$.

(2)   Next, a formal system **S** is said to be syntactically complete when, for any given formula α, either **S** proves α, or it proves $\neg\alpha$. When a system proves the negation of a formula, it is also claimed that the system *refutes* or *disproves* the formula. A formula which is either provable or refutable (disprovable) within a system **S** is said to be *formally decidable* in **S**. A syntactically complete formal system, therefore, is a system that, so to speak, can "make up its mind" on any formula of its underlying formal language: for any such formula α, the system will either "assert" (prove) α or "deny" it (prove α's negation). This kind of completeness is labeled as "syntactic," because it is defined by referring only to such syntactic notions as *theorem* and *proof*.

(3)   Third, (semantic) soundness or correctness is distinct from syntactic consistency (even though the latter sometimes is also referred to as a "soundness property"), for it is specified by reference to the notion of truth. A (semantically) sound system is a formal system that proves only truths: it is never the case that **S** proves α, when α is a false formula.

(4) Finally, a formal system **S** is said to be semantically complete when it proves *all* the true formulas, that is, it is not the case that some true formula is not a theorem of **S**.

Soundness and semantic completeness require that formulas be interpreted: it is only after a formula has been interpreted – that is, a meaning has been assigned to its constituents – that it makes sense to wonder whether it's true or false. And since, as we shall see, sometimes a formula can be true in some interpretation and false in some other, the interpretation at issue needs to be specified explicitly. When one claims that a given formula is true *simpliciter*, with no further qualification, one usually has in mind a privileged or "standard" interpretation. In the case of a formal system for arithmetic (I shall provide a detailed characterization of one such system in the following chapter), the standard interpretation has it that its formulas talk of natural numbers. Therefore, to claim that a formal system of this kind is sound is to say that it proves only arithmetically true formulas, that is, arithmetical truths, and to claim that it is semantically complete is to say that it proves all the arithmetically true formulas (expressible in the language).

The syntactic and semantic couples of properties are variously interrelated. A relation we are interested in now is the following: soundness is a *stronger* property than (syntactic) consistency. This means that the former entails the latter, but the latter doesn't entail the former: a sound formal system is consistent, but a consistent formal system can be unsound. The reason is fairly simple. An inconsistent system has as theorems both a formula and its negation, that is, a contradiction. Now a contradiction is not true under any interpretation.[9] Consequently, any inconsistent formal system is certainly unsound – just as a man who contradicts himself has certainly made some false claim. But a formal system can prove false things (or things that are false under a certain interpretation) even though it is syntactically consistent, that is, it does not prove contradictory statements, just like a consistent liar – a man who makes false claims, but without contradicting himself.

Keeping this fourfold terminology in mind, let us consider a formal system **S** on a suitable formal language **L**, which we take to be (semantically) sound or correct, that is, such that it proves only true formulas.

---

[9] Unless one is a fan of *strong paraconsistency* – a position I shall take into account in the second part of the book.

Let's assume that we can somehow express in the formalized language a self-referential statement of a certain kind: a sentence – call it G_S – similar to the Liar, but with the important difference that it claims of itself, not to be false (or untrue), but to be *unprovable* (in S):

(G_S) G_S is not provable in S.

As we know, the various Liar sentences produce contradictions. This happens when one begins to reason on the truth value of such statements as "This sentence is false," or "This sentence is untrue," and so on. Now, G_S is something like "This sentence is unprovable (in S)," and we can consider whether it is actually unprovable in S or not (Gödel himself explicitly stressed the proximity of his self-referential sentence to the Liar paradox).[10] Suppose G_S is provable. Then, given what it says, it is false, since it claims not to be provable. This would entail that the formal system S is *un*sound: it proves a false sentence, namely G_S. Therefore, if S is sound, as we have assumed it to be, G_S is not provable in it.

But if G_S is not provable, then G_S is what it claims to be; therefore, it's a true sentence. Then S is a *semantically incomplete* formal system: there exists a true sentence G_S which S cannot prove, i.e., which is not a theorem of S. Furthermore, since G_S is true, its formal negation $\neg G_S$ shall be false, given that the negation of a sentence is true if and only if the sentence is false, and vice versa. When all's said and done, neither G_S nor $\neg G_S$ is provable in S, and S turns out to be *syntactically* incomplete too: there exists a sentence such that neither it nor its negation is provable in S. So G_S is formally undecidable in S – which accounts for the title of Gödel's paper: "*On Formally Undecidable* Propositions ...."

What has just been presented is an informal semantic version of Gödel's First Incompleteness Theorem (I shall occasionally label Gödel's First Theorem as "G1" from now on). We can formulate G1 as follows:

(G1)   If S is a sound formal system, capable of expressing a certain amount of arithmetic, then there is a sentence G_S, expressed in the language L of the system, such that G_S is undecidable in S, that is, neither provable nor refutable in it.

---

[10] See Gödel (1931), p. 19.

Sentences of the kind G_S are usually called *Gödel sentences* (*of* a given formal system).

The (draft of a) proof of G1 just presented includes the claim "If S is sound, then G_S is not provable in S." We can weaken the antecedent of this conditional into: "If S is *consistent*, then G_S is not provable in S." Roughly, this hinges on the fact that if G_S were provable in S, then this proof would indicate that G_S is provable, and this last statement is nothing but $\neg G_S$ (for since G_S claims "G_S is not provable in S," then $\neg G_S$ claims "It is not the case that G_S is not provable in S"). Therefore, both G_S and $\neg G_S$ would be provable, and the system would be inconsistent.

Now, one can show that the phrase "If S is consistent, then G_S is not provable in S" is itself provable in S (even though Gödel didn't carry out the proof in full detail in his original paper – this is a point I will return to). And the consequent of this conditional is nothing but G_S itself, which claims precisely that G_S is not provable in S. So the phrase is equivalent to: "If S is consistent, then G_S." Now we rapidly obtain Gödel's Second Incompleteness Theorem (which I shall occasionally label "G2" in the following). In this "semantic" version, it looks like this:

(G2)   If S is a sound formal system, capable of expressing a certain amount of arithmetic, then S cannot prove its own consistency.

For let us suppose that S can prove its own consistency, i.e., that one can provide within S a proof of the claim "S is consistent." Then we would have a proof of the antecedent of the aforementioned conditional, that is, "If S is consistent, then G_S." By applying *modus ponens*, then, we could prove G_S. However, this has been ruled out by the *First* Theorem, which tells us precisely that G_S is not provable in S (so the Second Theorem is, in fact, a corollary of the First).

5   ... and some provisos

Arguably, it is this Second Theorem that spells trouble for Hilbert's Program. The key thought here is that Hilbert wanted to prove the consistency of formalized arithmetic only by finitary means. Now, one can certainly expect the metamathematical finitary reasoning employed in such a proof to be formalizable in a system capable of expressing arithmetic, as **S** has been assumed to be. But such a proof would be rendered impossible by G2, according to which a consistency proof for **S** cannot be carried out within **S**. Actually, as we shall see in the following, establishing the precise impact of G2 on Hilbert's Program is quite a subtle and delicate issue. Gödel himself didn't believe he had definitely crushed Hilbert's foundational ambitions. However, we shall deal with this issue too, little by little. The explanation of G1 and G2 just provided aims only at establishing an initial contact with Gödel's Theorem – specifically, (a) some things have to be made more precise, (b) some other things need to be slightly fixed, and (c) yet other things are probably difficult to grasp at this point, and dealing with them in full detail is a bit complicated. These will be the main tasks of the following chapters.

(a)   Among the things that have to be made more precise is the idea that a formal system can "express a certain amount of arithmetic." One has to specify what this certain amount amounts to (we shall also see that the "certain amount" required to prove the First Theorem is not exactly the same as the "certain amount" needed to prove the Second). One should also relate a more accurate story on what is meant by the claim that a system can "express" a portion of arithmetic.

(b)   Among the fixable things is the fact that, in the draft proof of the Theorems, we have assumed the system **S** to be sound, that is, such that it proves only true sentences. And we have carried out an informal reasoning to show that, under this assumption, G_S turns out to be true.[11] We have exploited, that is, the semantic concept *par excellence*: the concept of truth. More generally, we have resorted to the interpretation of the system **S** as formalizing certain notions. Now, in the actual proof of the Incompleteness Theorems carried out by Gödel in the main sections of his paper, he wanted "to replace the second of the assumptions just mentioned [namely, that every provable formula is true] by a purely formal and much weaker one."[12] Gödel knew very well that a formalist, or anyone subscribing to the viewpoint of Hilbert's metamathematics, would have found the resort to the notion of truth

---

[11] Also the claim that the Gödel sentence is *true* requires qualification; but we shall develop this point, too, step by step.
[12] Gödel (1931), p. 19.

and to the interpretation of the formal system rather unconvincing. In order to meet the formalists' qualms on truth, therefore, Gödel replaced the semantic hypothesis with a syntactic or metamathematical assumption on the formal system he was working with. Since, as we shall see, such an assumption is actually weaker than the assumption of soundness, Gödel's result is correspondingly stronger.

(c)   Third, among the things a careful reader might find initially quite difficult to grasp is the following: how can a sentence of a formal system for arithmetic *talk* (in any deliberative sense) *of itself*, thereby constituting a self-referential circle? To begin with, G_S is not written down in a formal logical language at all: it is a sentence phrased in (more or less) ordinary English. Now, building self-referential statements in ordinary English is not that difficult (this is why the Liar paradox has been around for 2,500 years). It is much more difficult to understand how a sentence formulated in a certain artificial language **L**, within a system **S** built in order to formalize some mathematics, can have self-referential features. A formal system for arithmetic, if one actually has to interpret it, is expected to talk about numbers and their properties, not about bits of language like sentences and theorems. As such, natural numbers are neither sentences nor sentence constituents.

For the same reason, it is difficult to understand how a sentence belonging to the language of a formal system **S** may talk of the consistency of that system, that is, may be (viewed as) claiming "**S** is consistent." Consistency, provability, etc. are, again, syntactic properties: they apply to formal systems taken as arrangements of linguistic objects. Once more, these do not appear to be numerical properties, such as the property of being even, or the property of being divisible only by itself and by one. Much of the cleverness in Gödel's result consisted precisely in making a marvelously innovative use of mathematics in order to overcome this difficulty. Gödel exploited the potentialities of the formalism to show that formalized arithmetic can achieve a measure of self reference: it can "talk of" itself as a linguistic system of formulas, axioms, and theorems, precisely *because* it can talk about natural numbers. To put it somewhat figuratively, such a formal system can carry out a certain amount of "self-analysis."[13] To understand how Gödel managed to do this, we shall now begin to play serious music.

[13]   "It somewhat resembles a self-conscious organism" (Smullyan (1992), p. 2).

3

Gödelization, or Say It with Numbers!

On October 7, 1930, Gödel gave the first public exposition of his Theorem. This happened during a congress held in Königsberg, to which some of the most influential logicians and mathematicians of the time had been invited: the intuitionist Arendt Heyting, to recount Brouwer's ideas; the Wittgenstein devotee Friederich Weismann; von Neumann, who talked of Hilbert's formalist project; and the logical positivist Rudolf Carnap, leading figure of the prominent Vienna Circle. Gödel wasn't one of the big guys: he gave a 20 minute talk and, contrary to what has been said and written, he did not produce any cataclysm. Gödel's approach was so strange and innovative that, it seems, nobody understood exactly what was going on.[1]

In the 1931 paper, Gödel focused on a specific formal system for arithmetic, which he labeled **P**. This was essentially a system obtained via some minor modifications from that included in Russell and Whitehead's *Principia mathematica*. Over the years, logicians have developed general abstract formulations of the Theorem. But Gödel was initially quite cautious about the extendibility of his results – not just because he was a cautious fellow, but also for a more technical and important reason, to be explained later. The approach I am adopting is quite close to the original Gödelian one: I will prove G1 and G2, not in some abstract version, as is often done today,[2] but by working on a definite formal system.

---

[1] With one exception: John von Neumann. Gödel had revealed only his First Theorem. After asking him for some explanations, von Neumann quickly managed to understand on his own that the Second Theorem followed – and with it, the impossibility of proving the consistency of formalized arithmetic within the theory. On this story, see Goldstein (2005), pp. 147 ff.

[2] See for instance Smullyan (1992); Boolos, Burgess, and Jeffrey (2002), Chs 17 and 18.

I shall, however, replace Gödel's notation and system with something more up-to-date. I'm going to play the original Gödelian symphony, but Fender guitars and synthesizers will replace Gödel's early-twentieth-century violins.

1   TNT

The theory to which Gödel's Theorem most classically applies is probably a formal system called *Peano Arithmetic*, labeled **PA**. I shall also work with this system, but I'm going to rename it *Typographical Number Theory*, and to label it **TNT**. The name comes from Hofstadter's *Gödel, Escher, Bach*, and it suits our needs particularly well. The reason is that we shall get used to "seeing things typographically": we shall learn to look at our formal system with a purely syntactic eye – with the eye, that is, of a formalist engaged in proof theory. We shall often look at formulas as mere strings of symbols, at rules of inference as mere instructions for symbol manipulation, and so on. In order to go beyond Hilbert's formalism, we have to begin by taking it seriously – which is exactly what Gödel did.

If the Gödel sentence is the main character of the book you are reading, the Typographical Number Theory is the character's mom. The logical symbols of **TNT** are just the usual ones of classical elementary logic (some of which have already been used since the first chapter to formulate some basic set-theoretic principles): the sign for the identity relation (=); the five logical connectives of conjunction ($\wedge$), disjunction ($\vee$), negation ($\neg$), the conditional ($\rightarrow$) and biconditional ($\leftrightarrow$); and the universal ($\forall$) and existential ($\exists$) quantifiers. Next, we need parentheses as auxiliary symbols. The language of **TNT** will also have a denumerable infinity of individual variables: $x, y, z, \ldots$ (I will occasionally use indices when more than three distinct variables are needed: $x_1, x_2, \ldots, x_n$).

As for the non-logical vocabulary, if you are familiar with elementary logic you should pay attention to the following point. When we study the language of first-order logic in general, we are usually authorized to introduce (perhaps contextually) all the individual, predicative, and function constants we like[3] – that is, respectively, the symbols denoting

---

[3]   As I have done in Berto (2007a).

individuals, properties (taken as sets of individuals or of *n*-tuples of individuals), and functions. We are allowed, that is, to be very prodigal with the descriptive vocabulary. But when one builds a specific formal system utilizing a formal language of this kind, one is supposed to follow an austere terminological diet: the only non-logical symbols around should be those that have been explicitly specified at the outset in the alphabet of the theory. Specifically, the non-logical vocabulary of **TNT** consists exclusively of the following four symbols: a one-place function symbol " "; an individual constant "**0**"; and two binary function symbols "**+**" and "**×**". The last three should look familiar from ordinary arithmetic: it is natural to see them as signifying respectively the number zero, addition, and multiplication.

However, this is precisely one of the cases in which we have to get used to seeing things typographically. We have, that is, to make a distinction between *informal*, intuitive arithmetic and its *formal* counterpart – and I shall use **boldfaced** symbols to stress the distinction. We have to keep in mind the difference between the (boldfaced) symbol **0**, which belongs to the vocabulary of Typographical Number Theory, and the natural number zero, which I'll keep indicating with 0 (not boldfaced). The latter is not a term of the **TNT** language, but an ordinary symbol utilized within (the fragment of) written English in which we express informal arithmetic. Analogously, we should make a distinction between the **TNT** symbols **+** and **×** (boldfaced), and the addition and multiplication operations of informal arithmetic, which I'll keep indicating with + and × (not boldfaced). Of course, **0** is taken as standing for the number zero, and **+** and **×** can be taken as standing for plus and times: this is the *intuitive interpretation* of those symbols. But, first, besides allowing us to keep the two kinds of symbols distinct, our double notation will help us to see the boldfaced signs purely syntactically when needed – to look at them as mere uninterpreted symbols. Second, the reading of (the boldfaced) **+** as the sum, etc., is at present only an intuitive association. In the following, we will fine-tune a precise notion of *formal representation* of intuitive arithmetical notions by (strings of) **TNT** symbols; only afterwards shall we be allowed to claim that that symbol **+** actually *represents* addition in Typographical Number Theory.

Let us come now to the least familiar of the symbols, that is, the apostrophe '. Its intuitive reading is that it stands for the (*immediate*) *successor* function (you should recall the notion of successor of a number from the first chapter). But the main "typographic" work done by ' consists in its allowing us to build *closed terms*, that is, terms that work like names, by repeatedly applying it to the (proper) name **0**. Within the **TNT** language, that is, we can have such closed terms as **0'**, **0"**, **0"'**, etc., obtained by having *n* occurrences of the apostrophe follow the symbol **0**. These terms are called the *numerals* of our Typographical Number Theory.

The intuitive reading of the numerals is that they denote natural numbers – exactly one number for each numeral. Thus, the official **TNT** name of number *n* is obtained by writing down *n* apostrophes after the **0**: number two's name is **0"**, which is quite intuitive, given that we can univocally describe that number as "the successor of the successor of zero"; number three's name is **0"'**, given that three is univocally describable as "the successor of the successor of the successor of zero"; and the name of the number 1,470,515 is the numeral obtained by having 1,470,515 apostrophes follow the **0**.

It is easily seen that, if we had to write down extensively the **TNT** numerals for big numbers, the notation would turn out to be rather unmanageable. To make our life easier, I will often indicate with a (boldfaced) **n** the numeral naming the natural number *n*. Thus, instead of writing down the numerals for the numbers two, three, …, etc. as **0"**, **0"'**, …, etc., I'll simply note them down as **2**, **3**, …, etc. (and, should we need the numeral for 1,470,515, instead of writing 1,470,515 apostrophes after the **0**, I would simply write **1,470,515**). Keep in mind, however, that this is just an informal expedient: the official numerals of the formal language remain defined as above.

2 The arithmetical axioms of TNT and the "standard model" ℕ

Keeping in mind all these annotations on the **TNT** vocabulary,[4] let us now come to the formal system in which the theory actually consists. We begin by taking on board some version or other of the classical

---

[4] And taking as specified the usual "grammatical" rules of well-formedness for terms and formulas of the logical language, which are a bit boring (those who have attended the famous course in basic logic presupposed by this book should have a precise idea of what they look like).

predicate calculus with identity, such as those we find in textbooks of first-order logic.[5] We obtain the full-fledged **TNT** by adding to the logic the following non-logical axioms:

(TNT1) $\forall x(x' \neq 0)$
(TNT2) $\forall x \forall y(x' = y' \to x = y)$
(TNT3) $\forall x(x + \mathbf{0} = x)$
(TNT4) $\forall x \forall y(x + y' = (x + y)')$
(TNT5) $\forall x(x \times \mathbf{0} = \mathbf{0})$
(TNT6) $\forall x \forall y(x \times y' = (x \times y) + x)$
(TNT7) $\alpha[x/0] \to (\forall x(\alpha[x] \to \alpha[x/x']) \to \forall x \alpha[x])$.

These axioms are formulated with the aim of providing a formal recapture of Peano axioms. In fact, in contemporary mathematical logic two big families of formal systems are taken into account. On the one hand we have *abstract* formal systems, in which the axioms themselves determine the fundamental properties of the notions at issue. When we derive consequences from the axioms, i.e., we derive the theorems of such formal systems, we are actually analyzing the properties and features of these notions.[6] When one builds the **TNT** system, however, one already has (or believes one has) in mind some structure or, as logicians say, some *model*: the model, let us say, consisting of the natural numbers, together with the operations on them and relations between them, which we began to learn at elementary school. This structure is what we want to capture by means of the formal theory.

To begin with, something more should be said on what models are in contemporary logic. Models are just (mathematical representations

---

[5]   Any version will do the job, for they are all provably equivalent: any system of logical calculus expressing first-order classical logic is sound and complete, in the sense that it allows one to derive all and only the (classical) logical consequences of the premises one assumes. So whether we take a Hilbert–Frege system with logical axioms (possibly given by means of schemes with metavariables) and at least one rule of inference (usually *modus ponens*), or a natural deduction calculus without logical axioms and supplied only with introduction-elimination rules, or yet some other system, this really makes little difference for our purposes.

[6]   This happens in some algebraic formal theories, such as group theories, etc. Such a purely formal account would have been approved by Hilbert, who (being criticized for this by Frege) took the axioms of formal systems as implicit definitions.

of) ontological structures in which formal theories are interpreted, and are usually phrased in set-theoretic terms. In the simplest cases, a model is something like an ordered couple – say <U, *i*>. U is some non-empty set, usually called the *domain* of the structure, and taken as the set over which the individual variables of (the language of) the theory range. When we meet such expressions as "for some *x*," "for all *x*," in the context of the language of a formal theory, what is meant, given an interpretation of the language in the relevant model, is "for all objects in the domain," "for some object in the domain." Classic textbooks also call the domain "universe of discourse," for it is the totality of the things the theory is supposed to talk about, in the given interpretation.

Next, *i* is an *interpretation function*. This is a function assigning meanings (denotations) to expressions of the formal language: individuals in the domain to individual constants; functions defined on the domain to function symbols; properties and relations (that is, sets of individuals and of *n*-tuples of individuals in the domain) to predicate constants. Logical terminology, in fact, is not completely standardized. Sometimes people make a distinction between a structure, or *frame*, and a model (the model being taken as the frame plus the interpretation function). Sometimes it is claimed that a given sentence (or a given formal theory) *has a model* in a structure when the structure makes the sentence (the whole theory) true. Equivalently, it is sometimes said that that the axioms and theorems of the theory are *satisfied* in (or by) the structure ("truth" and "satisfaction" are actually taken as naming distinct notions, especially after Tarski's work in formal semantics, but we need not enter into the details here).

Now, on the one hand we should learn to look at the **TNT** purely typographically – to see its axioms as strings of symbols, etc., as prescribed by the metamathematical approach. On the other hand, though, such axioms are not picked out at random. One wants + and × to actually denote the addition and multiplication operations, and the numerals to designate precisely the natural numbers; and, above all, one expects the variables of the language, *x*, *y*, ..., to vary precisely only on the familiar naturals our teacher told us about at school. Such a structure (the "intuitive interpretation") is called the *standard model* of **TNT** – call it ℕ – and its domain is precisely the set – call it N – of natural numbers. The interpretation of our **TNT** in the standard model is therefore called the standard interpretation. Within it, all the axioms

(TNT1)–(TNT7) hold. Gödel's Theorem will entail that things are not so straightforward for our theory: as we shall see in detail, the "capture" of the standard model attainable by **TNT** is inevitably defective. For the time being, however, let us stick to our *bona fide* intuitions and assume that (TNT1)–(TNT7) provide a satisfactory translation of Peano's principles.

Should we suspect at the outset the translation to be misguided, for instance, on the basis of the fact that the **TNT** axioms are seven, whereas the informal Peano axioms are five? Let us have another look at the informal principles:

(P1)   Zero is a number.
(P2)   The successor of any number is a number.
(P3)   Zero is not the successor of any number.
(P4)   Any two numbers with the same successor are the same number.
(P5)   Any property of zero that is also a property of the successor of any number having it is a property of all numbers.

(TNT1) and (TNT2) are the formal counterparts of (P3) and (P4). The main problem with the formalization comes from (P5), that is, the induction axiom. The point is that (P5) is phrased by saying "any property," that is, by quantifying on (talking in general of) all the properties of natural numbers. However, first-order logical languages don't allow one to say such things as "all properties," or "some properties." We have at our disposal individual variables that can be bound by quantifiers, so we can quantify on individuals ("for all x" and "for some x" mean "for all individuals in the domain" and "for some individuals in the domain"). But properties are beyond reach for the first-order quantifiers, and there are no predicate variables to be bound. We cannot talk in general of properties, or make such claims as "Napoleon had all the qualities of a great general." Logical languages allowing quantification on properties are usually called *second-order* languages, whereas our elementary language is called a *first-order* one (a terminology which has already been hinted at).

Now, when Peano formulated his axioms the distinction between first- and second-order logics and languages was not altogether clear. But if we wanted to translate (P5) literally, we would clearly need the expressive power of a second-order language. If one doesn't want to dispense with the induction axiom,[7] some trick is needed – and (TNT7) provides it. In the place of the property variables we cannot use, we find in (TNT7) the expression "α[x]". Such an expression, which we have already met in the first chapter, is a metavariable for formulas: a placeholder for any given object language formula with (one or more occurrences of) a free variable *x*. But metavariables themselves are not symbols that belong to the formal language of **TNT**. (TNT7), in fact, is an *axiom scheme*: a uniform substitution of α[x] with a given formula of the object language with (one or more occurrences of) a free variable delivers a full-fledged axiom. It is as if we were claiming: "Any formula with such-and-such a form – that is, any formula obtained by substituting α[x] with a given formula …, etc., etc. – is a **TNT** axiom."

How many axioms are present in our Typographical Number Theory, in the end? We have infinitely many of them, for the scheme (TNT7) originates an axiom for any **TNT** formula with a free variable, and we can potentially build infinitely many such formulas within the formal language.[8] One of the technical upshots of the situation is the following: we could define addition and multiplication in a second-order language, which is why we need no special principles for them among the informal Peano axioms. In a first-order formal language, however, we have to treat + and × as primitive, non-defined symbols, and we assume the axioms (TNT3)–(TNT4) and (TNT5)–(TNT6) respectively, in order to regulate their behavior.

3   The Fundamental Property of formal systems

Once the formal system was specified, Gödel made his first inspired move: he introduced what we nowadays call after him *Gödelization*,

---

[7]   And there are formal systems for arithmetic, weaker than **TNT**, that do without it – and despite being so weakened, still fall under Gödel's Incompleteness Theorem. One such system is *Robinson Arithmetic*, usually called **Q** (sometimes **R**). We will meet this formal system later.

[8]   This is a denumerable infinity, because the well-formed expressions of the **TNT** language are at most denumerable. On the other hand, the properties of natural numbers are all the subsets of the set N of naturals, so by Cantor's Theorem they are more than denumerable. This discrepancy has important consequences, as we will learn in the following.

or *Gödel numbering*. The insight as such is beautifully simple. Consider the Typographical Number Theory as I have just specified it, the way a formalist requires us to: take it as a system of (strings of) symbols, *mere symbols*, dispensing with their interpretation, their "intended reading," etc. Such a typographical way of looking at things, I have already said, is difficult, for we are naturally inclined to assign to symbols whose shape looks familiar the meanings we would normally ascribe them: for instance, one has the tendency to read the (boldfaced) symbol + as signifying addition. Let us make an effort, then, and look at **TNT** with the "eyes of the pure syntax," as Gödel suggests to us:

> The formulas of a formal system (we restrict ourselves here to the system [**TNT**][9]) in outward appearance are finite sequences of primitive signs (variables, logical constants, and parentheses or punctuation dots), and it is easy to state with complete precision *which* sequences of primitive signs are meaningful formulas and which are not. Similarly, proofs, from a logical point of view, are nothing but finite sequences of formulas (with certain specifiable properties).[10]

Now, even though Gödel had a specific system in his sights, his considerations hold for all of them. *Any* formal system **S** on a given formal language **L** is constituted by a denumerable set of discrete, well-determined objects.[11]

(1) First, we have the explicitly specified primitive symbols of the language **L** (logical symbols, such as connectives and quantifiers; non-logical symbols, such as variables and predicate constants; and auxiliary symbols, such as parentheses), which are either finite in number, or infinitely many (as happens in ordinary first-order formal systems, which include an infinity of variables $x, y, \ldots$) but denumerably so.

(2) Second, the rules of well-formedness for **L**, that is, the rules determining the syntax of the language, allow one to build terms and formulas out of the symbols, which in their turn are finite strings of symbols obtained by concatenating them (that is, by writing them one after the

---

[9] Recall that in the original text Gödel referred to (a slightly modified version of) the formal system of *Principia mathematica*.

[10] Gödel (1931), pp. 17–18.

[11] On the possibility of considering non-enumerable languages, though, see e.g. Boolos, Burgess, and Jeffrey (2002), pp. 162–3.

other) in certain ways. And the rules leave no room for qualms on the admissible ways. This is one of the decisive differences between any ordinary language and a formal logical one: the grammar and syntax of the latter are specified in a completely rigorous way, once and for ever. Given any sequence of symbols of **L**, the rules must allow us to decide, in a finite number of steps and in a completely effective way, whether the sequence is a well-formed or an "ungrammatical" one. To put it otherwise, the syntactic property of being a well-formed formula of (the language of) the system, or, equivalently, the set of well-formed formulas, has to be decidable, in the sense of the intuitive notion of decidable set we have met in the first chapter. The rules of well-formedness provide an algorithmic, effective procedure to decide in a finite number of steps whether a given string of symbols is a well-formed formula of the language.[12]

In particular, we can also establish in a finite number of mechanical steps whether a (well-formed) formula is an *axiom* of the system. The procedure is quite obvious for systems with a finite number of axioms – one just has to check through the list: if the formula appears in it, it is an axiom; otherwise, it isn't. But also in the case of systems with infinitely many axioms, such as our Typographical Number Theory, we can effectively decide, for instance, whether a formula is an instance of the induction scheme (TNT7), that is, whether it results from (TNT7) after the appropriate uniform substitutions have been performed. The syntactic property of being an axiom of the system, that is, the set of axioms, is itself decidable. A formal theory is said to be *axiomatized* in a strict sense only if the set of its axioms is decidable (and is said to be *finitely axiomatized* when such a set is a finite one).

(3)   Third, the formal proofs carried out within the formal system **S** are in their turn finite sequences of formulas. Typically, in the context of axiomatic systems each formula in the proof, as we have already anticipated, either is an axiom, or has been obtained from the preceding formulas by means of the logical rules of inference (such as *modus ponens*, for instance), and therefore is a theorem.[13]

---

[12]   I will most often just talk of "formulas," meaning the well-formed ones.

[13]   There are more complicated formal accounts, in which one is allowed to introduce formulas that are not axioms but "hypothetical" assumptions, and to have rules that discharge some assumptions (e.g. as in Berto (2007a), Ch. 3). These formal complications change nothing, though, in the general metamathematical status of formal proofs as syntactic objects, which is what we are interested in now.

From a purely syntactic viewpoint, inference rules are "typographical" transformation rules: they simply allow one to turn formulas, that is, finite sequences of symbols, into other formulas (to get an intuitive grip, we can actually think of the physical process of erasing, adding, and replacing symbols within formulas). And also in this case, when facing a (an alleged) formal proof in **S**, we can always decide in an effective way whether it is actually so – whether there are mistakes in the application of the rules, etc. Also the syntactic property of being a formal proof (in the given system **S**), that is, the set of formal proofs, is decidable.

As a matter of fact, these are essential requisites for having a real *formal system* – as opposed, let us say, to a "formalized theory" in a broad sense. To qualify something as a full-fledged formal system, logicians will not be content with a translation manual allowing a paraphrase from an informal theory into a symbolic logical language. It is required that axioms and rules of inference be defined in such a way as to satisfy what we may call the *Fundamental Property of formal systems*: **S** is a formal system only if *the set of its theorems is (computably) enumerable*. This follows from the fact that the axioms of a system have to form a decidable set, and the set of proofs is decidable too. To enumerate the theorems of the system (that is, to obtain them mechanically, one after the other) one has to check through the sequences of formulas, decide for any given sequence whether it counts as a proof, and, for each sequence that qualifies as one, pick the last line, that is, the respective theorem.

The point of the request put forth by claiming that a real formal system has to satisfy the Fundamental Property, or that the notion of proof in a real formal system has to be decidable, is that if we had to solve *further* mathematical problems in order to come to a decision on whether a given sequence of formulas is in fact a proof within a system **S**, we would need a further system of rules and principles – say **S₁** – to prove that an alleged proof in **S** is really such. And we would find ourselves involved in a dangerous regress.[14]

---

[14]   Let us listen to the authoritative logician Alonzo Church on this: "Consider the situation which arises if the notion of proof is non-effective. There is then no certain means by which, when a sequence of formulas has been put forward as a proof, the auditor may determine whether it is in fact a proof. Therefore he may fairly demand a proof,

Precisely because of these restrictions, the general notion of a formal system is intimately connected to those of decidable set (property, relation) and computable function. Gödel had already outlined the extendibility of his Theorem to different formal systems in his 1931 paper. However, to formulate his incompleteness results in the most general way he preferred to wait until a fully rigorous specification of the notion of formal system became available, as we shall see, by means of what he considered a satisfactory general theory of computability.

4   The Gödel numbering ...

When engaged in a "purely typographical" exploration of a formal system and of its formal language, as requested by Hilbert's metamathematical approach, we are dealing with a denumerable set of well-defined syntactic objects. This means that we can represent these objects by means of natural numbers, assigning a number to each object, distinct numbers for distinct objects. This is the core of the ingenious Gödelization procedure. It consists in specifying a function, say g, that assigns a single natural number to each symbol, term, formula, and sequence of formulas of the system, in a univocal and effective way. Given an expression (symbol, sequence of symbols, sequence of formulas) ε of the language of the system, the function maps it to a number $g(\varepsilon)$, called its *Gödel number*.

There are many different Gödel numberings around nowadays, that is, many different ways to assign natural numbers to expressions of formal systems. Any such assignment, however, has to fulfill the two following conditions: (1) it has to assign different numbers to different expressions (so it must never happen that the same number is assigned, say, to a formula and to a proof, or to two distinct formulas); and (2) the procedure has to be fully algorithmic and mechanical. This means that (a) for any given expression ε, it must be possible to compute its Gödel

---

in any given case, that the sequence of formulas put forward is a proof; and until the supplementary proof is provided, he may refuse to be convinced that the alleged theorem is proved. This supplementary proof ought to be regarded, it seems, as part of the whole proof of the theorem" (Church (1956), p. 53).

number $g(\varepsilon)$; and (b) for any given natural number n, it must be possible to decide whether it is the Gödel number of some expression of the language, that is, whether there exists an expression ε such that $n = g(\varepsilon)$; and in this case, it must also be possible to determine precisely which expression ε is.[15] Since both the function g and its inverse g^{-1} are computable, we can always move back and forth in an effective way between linguistic expressions and the corresponding numbers.

The move (a) from linguistic expressions to numbers is called *encoding*; the move (b) from numbers to linguistic expressions is called *decoding*; and the Gödel number of an expression is also called its *code*. The procedure resembles that of the coded messages used during wars – secret intelligence activities – but also that in childlike games. Consider the following approximate example. Suppose I want to transmit to my girlfriend some information during our English class, without being seen by the professor. I could use a simple encoding by pairing the space and the letters of the English alphabet to the initial numbers in the sequence of the naturals, following the normal lexicographic ordering, as follows:

	A	B	C	D	E	F	G	H	I	J	K	L	M
0	1	2	3	4	5	6	7	8	9	10	11	12	13

N	O	P	Q	R	S	T	U	V	W	X	Y	Z	
14	15	16	17	18	19	20	21	22	23	24	25	26	

Then, I could encode whole sentences as sequences of numbers corresponding to the letters they are made of, using the zero to signal the space between one word (that is, one finite sequence of letters) and another. In order to say to my girlfriend, for instance:

COME OUT NOW

I could write on a sheet:

3 15 13 5 0 15 21 20 0 14 15 23

---

[15] In some Gödel numberings, each natural number is the Gödel number of some expression (in this case, the function g is a bijection, e.g. as in Smullyan (1992)). In other versions, such as that provided in Gödel's original paper, not all natural numbers are Gödel numbers.

Being aware of the encoding at issue, my girlfriend could easily decode the message, moving back from numbers to letters, and therefore to sentences (which are finite sequences of letters and spaces), in an effective and mechanical way.

I have claimed that, provided the construction satisfies the afore-mentioned requirements, different Gödel numberings for the expressions of formal systems are available: theoretically speaking, it doesn't matter which one is chosen. However, to get to grips with the procedure I will provide a specific kind of Gödel numbering for our Typographical Number Theory, not too different from that employed by Gödel in his original paper.[16]

(1)   First, let us assign to each basic (logical, descriptive, or auxiliary) symbol s in the alphabet of the **TNT** language its Gödel number $g(s)$, by pairing each symbol with a distinct *odd* number as follows:

$g(\mathbf{0}) = 3$	$g(') = 5$	$g(+) = 7$	$g(\times) = 9$
$g(=) = 11$	$g(\neg) = 13$	$g(\wedge) = 15$	$g(\vee) = 17$
$g(\rightarrow) = 19$	$g(\leftrightarrow) = 21$	$g(\forall) = 23$	$g(\exists) = 25$
$g(\,(\,) = 27$	$g(\,)\,) = 29$	$g(x) = 31$	$g(y) = 33$
$g(z) = 35$	$g(x_1) = 37$	$g(x_2) = 39$	...

The fourteen constant symbols of the language of Typographical Number Theory – the four non-logical symbols, the eight logical ones (that is, the identity sign, the five logical operators, and the two quantifiers), and the two auxiliary symbols (the parentheses) – are assigned the odd numbers from 3 to 29. Next, we need odd numbers for all the infinitely many individual variables; but since there are infinitely many odd numbers greater than 29, we have enough of them.

(2)   One can virtually construct infinitely many well-formed formulas out of the vocabulary of **TNT**. How do we manage to assign a single natural number to each of them in a suitable way? Gödel ingeniously exploited the Fundamental Theorem of Arithmetic, or the Unique Prime-Factorization Theorem, which claims that every natural number (greater than 1) can be uniquely decomposed as a product of prime numbers.

---

[16]   This is taken from Palladino (2004), pp. 131–2 – an excellent introductory textbook on first-order formal systems and theories.

Therefore, given any finite sequence of numbers – say the sequence $k = m, n, o, p, q, \ldots$ – we can assign to k the following single number:

$$2^m \times 3^n \times 5^n \times 7^p \times 11^q \times \ldots$$

Now, any well-formed formula α in the language of **TNT**, as we know well by now, will always be a finite sequence of symbols of the formal vocabulary, built in accordance with the grammatical rules. Thus, if a given formula α consists, say, of the sequence of symbols $s_1, s_2, s_3, \ldots, s_n$, we can "sum it up" in a single Gödel number, that is, a product of precisely n powers: those whose bases consist of the primes listed in increasing order, and whose exponents are the Gödel numbers of the symbols $s_1, s_2, s_3, \ldots, s_n$:

$$g(\alpha) = 2^{g(s1)} \times 3^{g(s2)} \times 5^{g(s3)} \times \ldots \times p_n{}^{g(sn)}$$

where p_n is the nth in the increasing ordering of the primes. Given the structure of the Gödel numbering, we can therefore perform the encoding and decoding in an effective way in this case as well. When assigning codes to the basic symbols of the language of our Typographical Number Theory we had to make use of all the odd naturals; but the Gödel numbers of the formulas are all even, so no superposition can take place.

Let us look at a couple of simple examples of coding and decoding of formulas. Take the simplest **TNT** formula: $0 = 0$. The codes of the symbols it is made of are:

0	=	0
3	11	3

Its Gödel number, therefore, is $2^3 \times 3^{11} \times 5^3 = 177,147,000$. And we can go round the other way: given the number $177,147,000$ we can analyze it in a unique way into prime factors, as $2^3 \times 3^{11} \times 5^3$. The bases of the powers are the first three primes. Their exponents, 3, 11, and 3 again, are all odd and correspond precisely to the symbols **0**, =, and **0** again. We can therefore make a simple calculation and establish that $177,147,000$ is the Gödel number of the formula $0 = 0$.

Next, consider the formula $\exists x(x = y')$. The codes of the symbols it is made of are:

∃	x	(	x	=	y	'	)
25	31	27	31	11	33	5	29

Thus, its Gödel number is $2^{25} \times 3^{31} \times 5^{27} \times 7^{31} \times 11^{11} \times 13^{33} \times 17^5 \times 19^{29}$. The fact that this can overheat your pocket calculator is irrelevant: the important point is that, even when we are dealing with huge numbers, the encoding and decoding could in principle always be carried out in an effective and mechanical way.

(3)   Finally, the formal proofs carried out in **TNT** are finite sequences of formulas, that is, finite sequences of finite sequences of symbols. Thus, we can assign them Gödel numbers by exploiting again the prime decomposition stratagem. Given a sequence of formulas $\sigma = \alpha_1, \alpha_2, \alpha_3, \ldots, \alpha_n$, its unique Gödel number $g(\sigma)$ will be a product of n powers, having as bases the primes in increasing order, and as exponents the codes of the relevant formulas:

$$g(\sigma) = 2^{g(\alpha 1)} \times 3^{g(\alpha 2)} \times 5^{g(\alpha 3)} \times \ldots \times p_n{}^{g(\alpha n)}.$$

The codes of the formulas have odd numbers as exponents (those assigned to the symbols that the formulas are made of), whereas the codes of the sequences of formulas have as exponents even numbers (those assigned to the formulas that the sequences are made of), so they differ from all the preceding ones: there is no superposition in this case either.

Not every number is a Gödel number in this Gödelization. For instance, consider 200. Since it is even, it is not the code of a basic symbol. Its prime decomposition is $2^3 \times 5^2$. This does not have the required form, because the prime factor 3 is missing. So 200 is not the Gödel number of any expression of the language of **TNT**.

5   … and the arithmetization of syntax

The notions applying to expressions of the formal system, that is, to syntactic objects (such as the notion of formula and the notion of proof), correspond to arithmetical concepts applying to the Gödel numbers or codes of those objects. This produces a situation which has been called *arithmetization of syntax* (or of metamathematics). The terminology is easily understood. There exists an effective correspondence between syntactic properties and relations (that is, properties of and relations between expressions of the language of **TNT**) and arithmetic properties and relations of the respective codes. For instance, a given formula α is *derivable from* a formula β if and only if there exists some arithmetical relation between the respective Gödel numbers $g(\alpha)$ and $g(\beta)$. Notice that something of the sort would happen even if we encoded numerically non-linguistic objects. For instance,[17] when we go to the supermarket we are assigned a number in order to be waited on; and the non-mathematical fact that I, having been given the number 74, will be served after Ms Bellucci, who has been given the number 69, is mirrored in the fact that $69 < 74$, that is, in an arithmetical relation between two numbers.

Now since the Gödel function g (as well as its inverse) is effectively computable, decidable syntactic-metamathematical properties and relations concerning the formal language of the system (such as the property of being a variable, or those of being a well-formed formula, an axiom, or a formal proof) correspond to decidable arithmetical properties and relations (the property of being the Gödel number of a variable, or those of being the Gödel number of a well-formed formula, of an axiom, etc.). I have claimed that which Gödel numbering one chooses is largely conventional. But the non-negotiable point of the construction is that the syntactic properties of the strings that make them recognizable as strings of a certain kind correspond to decidable arithmetic properties of the respective codes. For instance, in the sample Gödel numbering I have given above, the property of being the Gödel number of a variable is just the property of being an odd number greater than 29.

We should keep in mind that the very same procedure can be applied to any formal system, for any such system will always consist of a language with a well-defined set of basic symbols, of well-formed formulas which are finite sequences of such symbols built according to precise grammatical rules, etc., etc. This explains the title of this chapter, "Say It with Numbers": we can always *talk about* syntactic notions, that is, about certain properties of and relations between linguistic expressions, by resorting to arithmetic, via the corresponding arithmetical properties of the respective numerical codes: *simplex sigillum veri*.

Gödel numberings and arithmetization are nowadays standard practice in mathematical logic and computability theory, and have a variety of applications. In order to grasp how Gödel put the procedure to work in his paper, though, we need to take a small mathematical detour; this we are about to do in the following chapter.

[17] This comes from Nagel and Newman (1958), pp. 80–1.

4

Bits of Recursive Arithmetic …

In his paper Gödel defined a series of 45 arithmetic functions-relations, called *recursive*.[1] The functions had already been used by Dedekind, Skolem, Hilbert, and Ackermann, but the sequence of definitions provided by Gödel has become canonical. In this chapter I will explain some basics of recursion theory. This might appear to be a detour from the main path we are following[2] but, as we shall discover later, it isn't. On the one hand, while in the business of obtaining his fundamental result, that is, the Incompleteness Theorem, Gödel provided an account of the theory of recursive functions, which is nowadays at the core of studies of the logical and mathematical phenomenon of computability. On the other hand, Gödel's definitions helped in establishing that fundamental result itself – as we shall see in the following chapter. We will be interested in recursive functions only in so far as they serve as means to understand our primary theoretical target.

1   Making algorithms precise

So far I have been talking of decidability for sets (properties and relations) with respect to sets of numbers, in Chapter 1, and with respect to sets of expressions of formal languages, in the previous chapter. But all was presented pretty informally and by relying on intuition. The reason is, as I have already stated, that the notion of algorithm – i.e., of

---

[1]   Gödel's 45 recursive functions do not exhaust the totality of such functions, as we shall see in the following.
[2]   In his paper, Gödel presented it as "a parenthetic consideration that for the present has nothing to do with the formal system" (Gödel (1931), p. 23).

a mechanical, deterministic procedure terminating after a finite number of steps, etc., etc. – is itself an intuitive one. The theory of recursive functions is a way of making the notion much more precise. We especially need to be more accurate when we aim at establishing that some sets are not decidable (or that some functions are not computable) by means of an algorithm. By resorting to the intuitive notion, it is not too difficult to show that some problem *is* solvable via algorithmic operations. But one of the most interesting results of formal theorizing on computability consists in the discovery that some problems are *not* algorithmically solvable; and to establish this in a fully accurate way we need to set precise boundaries for the realm of the computable.

One can enquire whether a set or a property is algorithmically decidable with respect to sets and properties of objects of various kinds, besides numbers. For instance, in the previous chapter it has been claimed that the property of being a well-formed formula of a given formal language **L**, and the property of being an axiom of a formal system **S** (that is, the set of well-formed formulas, the set of axioms), are decidable (on the basis of the grammatical rules of well-formedness, etc.). And these are properties or sets of linguistic expressions – pieces of language, not numbers. However, precisely the things we learnt in that chapter tell us that, if one can meaningfully ask questions about computability with respect to properties or sets of things of a certain kind, then such questions can always be reduced to arithmetical problems, that is, to problems concerning (properties and sets of) natural numbers. This is the most significant outcome of the fact that, via the Gödel numbering, we can "say it with numbers." If we can build an effective one-to-one correspondence between the elements of a set whose features are under investigation, and (a subset of) the naturals, then questions concerning the items in the set turn into corresponding arithmetical questions on the associated numbers: we encode the elements of the set via the naturals, and we posit the problem with reference to the codes themselves.

2   Bits of recursion theory[3]

Recursive functions are so called because their definition is by *recursion*, or by *induction*, starting from some basic or initial functions

---

[3] The exposition in this section draws on the classic Boolos, Burgess, and Jeffrey (2002), Ch. 6.

taken as intrinsically simple. The basic recursive functions are the following.

The *zero* function z, given any natural number x as argument, always has zero as the corresponding value: $z(x) = 0$. Its course of values is thus: $z(0) = 0, z(1) = 0, z(2) = 0, \ldots$ (a tedious function!)

The *successor* function s, given a number x as argument, assigns as value the next larger number, that is, its successor: $s(x) = x + 1$. Therefore: $s(0) = 1, s(1) = 2, s(2) = 3, \ldots$.

Then we have a bunch of *n*-ary functions p_i^n, called the *identity* or *projection* functions. There is one identity-projection function of one argument, p_1^1: for any number x, its value is the number itself: $p_1^1(x) = x$ (therefore: $p_1^1(1) = 1, p_1^1(2) = 2, p_1^1(3) = 3, \ldots$). Next, there are two identity functions of two arguments, p_1^2 and p_2^2: given any pair of numbers x and y as arguments, the former function always gives as value the first one, and the latter the second: $p_1^2(x, y) = x, p_2^2(x, y) = y$. Since the underlying idea is clear, we can easily generalize: for each positive integer n there are n identity functions, p_i^n, of n arguments, assigning the ith (the first, the second, ..., the nth) of such arguments as value: $p_i^n(x_1, \ldots, x_n) = x_i$, with $1 \le i \le n$. For instance, $p_2^3(2, 5, 1) = 5$, and $p_5^7(72, 4, 11, 128, 23, 8, 9) = 23$.

These initial or basic functions are manifestly computable in the intuitive sense: the procedure to compute their values for any argument(s) is algorithmic. For instance, given the successor function with argument 7, we add one to obtain the value: $s(7) = 7 + 1 = 8$; and given the identity function p_5^7 with the arguments $72, 4, 11, 128, 23, 8, 9$ (in that order), to obtain its value we just have to read through the list until we find the fifth number, that is, 23. The initial functions can be computed, as it were, "in one step."[4]

Now, the set of recursive functions is the set of all and only the functions that can be defined, starting with the basic functions, by means of some operations producing new functions from old ones, and which *preserve* computability. This means that, if some functions are computable, the functions defined by applying to them such operations are computable in their turn. The first two operations admitted are the following.

(1)   The *composition* or *substitution* operation is quite simple: given a function f of n arguments, and a series of functions $g_1, \ldots, g_n$ all of m arguments, one claims that the following function h of m arguments:

$$h(x_1, \ldots, x_m) = f(g_1(x_1, \ldots, x_m), \ldots, g_n(x_1, \ldots, x_m))$$

---

[4]   Ibid., p. 64.

has been obtained by composition or substitution from f and $g_1,\ldots,g_n$. If these functions are effectively computable, h is computable too: that is to say, composition preserves computability. The number of steps one has to perform to compute the value of h given the arguments $x_1,\ldots,x_m$ is the sum of the number of steps required to compute the value (say y_1) of $g_1(x_1,\ldots,x_m)$, the number of steps required to compute the value (say y_2) of $g_2(x_1,\ldots,x_m),\ldots$, and so on to the value y_n of $g_n(x_1,\ldots,x_m)$, plus the number of steps one needs to compute $f(y_1,\ldots,y_n)$.

(2)   The (*primitive*) *recursion* operation is the following: when f is a function of n arguments, and g is a function of $n + 2$ arguments, one claims that the function h of $n + 1$ arguments specified thus:

(i) $h(x_1,\ldots,x_n,0) = f(x_1,\ldots,x_n)$

(ii) $h(x_1,\ldots,x_n,s(x)) = g(x_1,\ldots,x_n,x,h(x_1,\ldots,x_n,x))$

has been obtained by (primitive) recursion from f and g.

This recursion is indeed a kind of mathematical induction. The idea is that (i) one defines the output of the function for the input zero; and (ii) based upon the definition of the output of the function for a given x, one defines the value for $s(x) = x + 1$, that is, for the next larger number. Now if f and g are computable functions, h is too: one just proceeds backwards with respect to the induction. Given clause (ii), the value of h in x can be reduced to the value of its predecessors, that is, $x - 1, x - 2$, etc. Once we reach zero, we get the final value by means of clause (i). Recursion, at the end of the day, preserves computability.

A function is said to be *primitive recursive* if and only if either it is one of the basic functions, or it has been defined from them by applying the composition and/or recursion operations a finite number of times. Many intuitively computable functions and operations of ordinary arithmetic can be shown to be primitive recursive. For instance, addition is primitive recursive. The recursive rules for addition are just the following two equations:

(+i) $x + 0 = x$

(+ii) $x + s(y) = s(x + y)$

and these conform to the recursion scheme given above. Just as we have defined (the rules for) addition via the successor function, so we can define multiplication by means of addition, since multiplication is repeated addition: 5×4 is just $5 + 5 + 5 + 5$. So we have the following equations:

(×i) $x \times 0 = 0$

(×ii) $x \times s(y) = x + (x \times y)$

and we can reduce the calculation of a product to calculations of sums. Next, since exponentiation is just repeated multiplication, it can be defined by means of the latter: 5^4 is just $5 \times 5 \times 5 \times 5$, so we can reduce exponentiation to the computation of products.

One can show that a lot of other arithmetic functions are primitive recursive (e.g., the factorial function, the predecessor, the difference between naturals, then any inverse function of a bijective primitive recursive function, etc., etc.). However, some intuitively computable functions are not primitive recursive (the first and best known is called the Ackermann function, for it was discovered by Ackermann in 1928). But even these fit within the broader concept of *general* recursive function. The set of general recursive functions (sometimes called recursive functions *simpliciter*) is the set of functions one obtains from the basic ones, not only via substitution and (primitive) recursion, but also by means of another operation, called *minimization*, or sometimes μ-recursion. Minimization is based upon the principle of the smallest number (that is: if we know that some natural number has a property P, we also know there is a smallest natural number with P). I shall skip this further development, though: on the one hand, because a grasp of the notion of primitive recursion already gives us a sufficiently precise idea of how the intuitive notion of computability is characterized within recursion theory (basically, the theory works by building new functions as constructions out of the basic functions, via the admissible procedures and the functions previously defined); and on the other hand, because for the proof of Gödel's Theorem one can take into account only the primitive recursive functions.[5]

---

[5]   It is worth mentioning, though, that many basic results within general recursion theory (e.g. the Normal Form Theorem, the s–m–n Theorem or Kleene's Recursion Theorem) are obtained by making essential use of the Gödelization procedure, applied to the recursive

3   Church's Thesis

Still, we are now (extremely!) interested in the following fact. All (primitive and general) recursive functions are effectively computable in the intuitive sense. The hypothesis that, conversely, all effectively computable functions are recursive is called *Church's Thesis*, for it was put forward by the American mathematician and logician Alonzo Church.[6] If Church's Thesis holds, the functions we take as computable in the intuitive sense *coincide* with the recursive ones. We must now understand (a) why this Thesis is important, and (b) what the claim that it's a *thesis* means.

(a)   The reason why Church's Thesis is important is that some functions studied in mathematics are not recursive. Now, I have already claimed that one cannot show rigorously that a function is not effectively computable, for effective computability is an intuitive notion and, in this sense, one with fuzzy borders. But one can rigorously demonstrate that some functions are not recursive: it is sufficient to show that they cannot be defined by starting with the basic functions and using only the operations allowed by recursion theory. Therefore, if Church's Thesis holds, such functions are simply not effectively computable. And from this one may infer that "logicians and mathematicians would be wasting their time looking for a set of instructions to compute the function."[7]

(b)   The reason why Church's is called a *thesis* is that it has not been rigorously proved and, in this sense, it is something like a "working hypothesis." Its plausibility can be attested inductively – this time not in the sense of mathematical induction, but "on the basis of particular confirming cases." The Thesis is corroborated by the number of intuitively computable functions commonly used by mathematicians, which can be defined within recursion theory. But Church's Thesis is believed by

---

functions themselves. Codes, that is, Gödel numbers (usually called *indexes*) are assigned to functions, and since the functions in their turn have natural numbers as their inputs, it is possible, for instance, to embed the code of a function as an input of another function. This is the analogue of what happens in any computer, in which instructions and computation procedures are numerically encoded in the same format as the data the procedures compute. I will come back to this analogy and its importance in a later chapter.

[6]   One usually refers to total functions. The hypothesis that all intuitively computable partial functions are (partial) recursive functions is sometimes called *extended* Church's Thesis.

[7]   Boolos, Burgess, and Jeffrey (2002), p. 71.

many to be destined to *remain* a thesis. The reason lies, again, in the fact that the notion of effectively computable function is a merely intuitive and somewhat fuzzy one. It is quite difficult to produce a completely rigorous proof of the equivalence between intuitively computable and recursive functions, precisely because one of the sides of the equivalence is not well-defined. Church's Thesis, thus, according to some looks a bit like a scientific conjecture, in the sense Karl Popper would give to the term "conjecture": something we might falsify (should we discover a computable but non-recursive function), but never definitely verify.[8]

Another reason why Church's Thesis is plausible, though, is that during the twentieth century various other formal characterizations of the intuitive notion of computability were independently proposed: among them, the theory of Markovian algorithms, and Alonzo Church's lambda calculus. I shall (quickly) deal only with one of these theories, that is, the theory of Turing machines, in the following. The relevant remark here is that these theories, though elaborated largely autonomously, have turned out to be, surprisingly, equivalent to each other in a precise sense: they all individuate the same set of functions. The as it were "extensional" equivalence between these theories corroborates each of them separately, and corroborates Church's Thesis too. The fact that theories built up in different ways and adopting diverse basic notions individuate the same group of functions suggests on the whole that *each* of them has captured the (single) authentic concept of effectively computable function in its own way.

4   The recursiveness of predicates, sets, properties, and relations

We can now introduce some more terminology. Given Church's Thesis, we can begin to use the expressions "computable function" and "(general) recursive function" interchangeably. But we can also allow ourselves some other synonymies. Even though in this chapter I have been talking in terms of functions, the discourse can be extended to predicates, sets, properties, and relations. From the first chapter, we know that each set has its characteristic function, that is, the function whose value is 1 if the argument (or the *n*-tuple of arguments) belongs to the set, 0 otherwise.

---

[8]   This was Kalmàr's (1959) position, for instance.

We also know that a set is decidable if and only if its characteristic function is computable. Correspondingly, a set is (primitive) *recursive* iff its characteristic function is. Church's Thesis, according to which all computable functions are recursive, entails that all decidable sets are recursive. And the terminology is easily generalized to properties (which can also be taken as sets, as we know well), n-ary relations (considered as sets of ordered n-tuples), and predicates (that is, the linguistic entities which denote properties and relations, therefore sets). We shall therefore have that a relation is effectively decidable if and only if it is recursive, etc. And we can begin to use as synonyms also "decidable set (property, relation, predicate)" and "recursive set (property, relation, predicate)."

Moreover, the intuitive notion of computably enumerable set (relation, etc.) which we met in the first chapter has its formal counterpart in recursion theory via the notion of a *semi-recursive* or *recursively enumerable* set. A set is recursive if and only if its characteristic function is. A recursively enumerable set can be characterized as a set which is the range of a recursive function undefined for some arguments.[9] And the formal counterpart of the fundamental idea that if both a set and its complement are semi-decidable, then the set is also decidable, is Kleene's Complementation Theorem:[10] if a set and its complement are both recursively enumerable, then that set is recursive.

Now, back to the 45 (primitive) recursive functions-relations defined by Gödel in his 1931 paper. These are functions and relations which are naturally at issue when one examines the syntax of a formal system via the Gödel numbering procedure. Gödel built the list very meticulously, and the first four Theorems in Gödel's paper deal with the general properties of the items in the list. But, as we shall see in the next chapter, it is with Theorem V that the main movement of the Gödelian symphony begins.

---

[9]   And the counterpart in recursion theory of the things I have been saying on computably enumerable sets that are not decidable is the following: recursively enumerable sets display a basic *asymmetry* between belonging and not belonging to them. If something is a member of a recursively enumerable set, this can be effectively ascertained, whereas if something is not a member, we may have no mechanical procedure to come to know it. So we have a merely *partial* algorithmic test to determine whether a given x is a member of a recursively enumerable set or not: we can effectively produce the members of the set, and hope that the guy we are looking for shows up sooner or later (see e.g. Bellotti, Moriconi, and Tesconi (2001), pp. 152 ff).

[10]   See Boolos, Burgess, and Jeffrey (2002), p. 82.

5

… And How It Is Represented in Typographical Number Theory

In this chapter, the circle closes around our Typographical Number Theory. The time has come to complete the exposition of the stratagem which allowed Gödel to circumvent the Hilbertian prescription to separate theory and metatheory – specifically, the prescription to the effect that the formalized Typographical Number Theory be kept utterly isolated from the metamathematical claims *on* the formal system itself (its language, the syntactic properties of its strings of symbols, and so on).

1 Introspection and representation

Chapter 3 has shown us how one can arithmetize the syntax of any given formal system via the Gödel numbering stratagem. By a suitable coding of the expressions (symbols, formulas, etc.) of a given system **S** on a formal language **L**, we can talk about the formal system and its language in terms of natural numbers, their properties and relations. As it turns out, arithmetization associates such syntactic notions as those of formula and proof with arithmetic concepts. Specifically, decidable properties of, relations between, and sets of linguistic expressions are paired with decidable arithmetic properties, relations, and sets: those that concern the respective codes or Gödel numbers.

In addition, from Chapter 4 we have a more precise characterization of *decidable* in terms of recursive functions-relations: given Church's Thesis, we can equivalently talk of recursive properties, relations, and sets – notions that receive a rigorous foundation within recursion theory. In particular, we can adopt the following terminology: given a formal system **S** on a language **L**, and a suitable Gödel numbering, we can claim that a set of expressions (symbols, formulas, etc.) in the language of **S** is recursive in a derivative sense, meaning that the set of codes of those expressions is recursive (and one can generalize to properties, relations, etc., in the usual way). Thus we say, for instance, that the set of sentences of the language of **S** (that is, of the closed formulas with no free variables), or the set of proofs in **S**, are recursive; and what we mean is that the sets of Gödel numbers of the sentences, or of the proofs, are recursive.

I have claimed that these outcomes of the arithmetization of syntax hold for any formal system whatsoever. It is part of the very notion of a formal system that the sets of its axioms and of its proofs be decidable. However, systems like our **TNT** have something that makes them very *special* in this respect. The arithmetization procedure allows us to talk of the expressions, formulas, and proofs of any given formal system (**TNT** included) in terms of natural numbers. But the **TNT** system has been built precisely in order to formally represent arithmetic. We can therefore expect some syntactic properties and relations concerning the system, and associated with arithmetical properties and relations via the Gödel numbering, to be somehow representable *within* our Typographical Number Theory itself. Things being so, **TNT** may be able to express significant portions of its own syntax. An informal way to describe the situation, which I have already hinted at, consists in claiming that **TNT** is capable of some "introspection," or "self-analysis."

That **TNT** may "talk about" its own syntax should already make us scent self-reference. What happens with systems such as **TNT** is that natural numbers play a twofold role. First, they can be seen as what the **TNT** sentences "officially talk" about. But second, because of the Gödel numbering the naturals are also codes paired to the expressions of the language of **TNT** itself. And this is how, in officially "talking about" numbers and their properties, **TNT** may turn out to "talk about" itself – about various features of the strings of symbols it consists of.

In fact, this is still a rough characterization of the situation – albeit one you can find in various textbook expositions of Gödel's Theorem. The (draft) theory of recursive functions in the previous chapter was presented in a largely informal way: it was delineated in ordinary English (even though I used some elementary mathematical notation with symbols for functions, variables, etc., to make my life easier), and not axiomatized in any sense. As such, it belonged to what has been previously called "intuitive" arithmetic. To understand more accurately what one means by claiming that **TNT** can "talk about its syntax" because it can "talk about" the recursive functions-relations, it is time to "make talk about 'talking about' precise"![1]

To put it otherwise: in order to rigorously establish how much of intuitive arithmetic can be formally *represented within* the **TNT** system, we have to provide a precise characterization of the notion of *representability*.

2   The representability of properties, relations, and functions ...

A k-ary relation R between natural numbers, as has been claimed several times, can be taken as a set of ordered k-tuples of naturals. Now, R is said to be *representable* within our **TNT** if and only if there is a formula $\alpha[x_1,\dots,x_k]$ of the **TNT** language with exactly k free variables $x_1,\dots,x_k$ such that, for each ordered k-tuple of numbers $<n_1,\dots,n_k>$,

(a)   If $<n_1,\dots,n_k> \in R$, then $\vdash_{\text{TNT}} \alpha[x_1/\mathbf{n}_1,\dots,x_k/\mathbf{n}_k]$

(b)   If $<n_1,\dots,n_k> \notin R$, then $\vdash_{\text{TNT}} \neg\alpha[x_1/\mathbf{n}_1,\dots,x_k/\mathbf{n}_k]$

where $\mathbf{n}$, recall, is the numeral for the number n. Thus, "$\alpha[x_1/\mathbf{n}_1,\dots,x_k/\mathbf{n}_k]$" indicates the formula obtained from $\alpha[x_1,\dots,x_k]$ by substituting the variable x_1 with the numeral $\mathbf{n}_1$, the variable x_2 with the numeral $\mathbf{n}_2,\dots$, etc.[2]

Clause (a) in the definition of representable relation, translated into English, tells us that, if R holds across the numbers $n_1,\dots,n_k$ (in this order), then the formula $\alpha[x_1/\mathbf{n}_1,\dots,x_k/\mathbf{n}_k]$ with the corresponding numerals is a theorem of **TNT**. And clause (b) tells us that, if R does not hold across

---

[1]   Boolos, Burgess, and Jeffrey (2002), p. 199.
[2]   And the syntactic operation of substituting, in a given formula with some free variables, these very variables with terms, should be familiar to those who have attended the famous course in basic logic presupposed here. I shall talk of substitution at length in the following.

$n_1, \ldots, n_k$ (in this order), then the negation of that formula is a **TNT** theorem. It is claimed, then, that the formula *represents* (sometimes, *numeralwise represents*)[3] the relation at issue.[4]

The properties of the naturals, that is, the sets of natural numbers, are just a particular case of this situation. One claims, that is, that a set - say M - of naturals is representable within **TNT** when there exists a formula $\alpha[x]$, with the sole variable x free, such that, for any natural n,

  (a) If $n \in M$, then $\vdash_{\text{TNT}} \alpha[x / \mathbf{n}]$

  (b) If $n \notin M$, then $\vdash_{\text{TNT}} \neg \alpha[x / \mathbf{n}]$.

Translating into English again, this means that (a) if n belongs to the set M (that is, it has the relevant property), then the formula $\alpha[x/\mathbf{n}]$ with the corresponding numeral is a **TNT** theorem; and (b) if n does not belong to M (it lacks the relevant property), then the negation of the formula is a **TNT** theorem. It is claimed, then, that the formula (numeralwise) represents the property or the set at issue.

Now that we have a precise definition of the representability of properties, sets, and relations of intuitive arithmetic within our **TNT**, we can look at a couple of examples. The equality relation between numbers is representable in **TNT**. This will be a set - say I - of ordered pairs, having as members all and only the pairs of numbers $<m, n>$ such that m is equal to n. The symbol of the **TNT** language whose task is to represent equality is, of course, the identity sign "=". Since it can be shown that, given two numbers m and n,

  (a)   If $< m, n > \in I$, then $\vdash_{\text{TNT}} \mathbf{m} = \mathbf{n}$

  (b)   If $< m, n > \notin I$, then $\vdash_{\text{TNT}} \neg(\mathbf{m} = \mathbf{n})$,

---

[3]   E.g. in Kleene (1952), p. 200.
[4]   If a k-ary relation R is such that

  $< n_1, \ldots, n_k > \, \in R$, if and only if $\vdash_{\text{TNT}} \alpha[x_1 / \mathbf{n}_1, \ldots, x_k / \mathbf{n}_k]$

then one claims that R is *semi-representable* or *weakly representable*. Representability and weak representability should be kept distinct. If a relation R is weakly representable, then the fact that $n_1, \ldots, n_k$ (in this order) are not in the relation R only entails that $\alpha[x_1/\mathbf{n}_1, \ldots, x_k/\mathbf{n}_k]$ is not a theorem of **TNT**. It does not entail, though, that $\neg \alpha[x_1/\mathbf{n}_1, \ldots, x_k/\mathbf{n}_k]$ *is* a theorem of **TNT**.

we can maintain that the equality relation between natural numbers is actually represented within **TNT** by the symbol "=".

The property of being even, that is, the set of even numbers, is also representable in **TNT**. It is represented by the formula $\exists y(y \times 2 = x)$ (recall that **2** is an abbreviation for **0"**), with the sole variable x free. This is because it can be proved that, if E is the set of even numbers, then for any given number n,

(a)   If $n \in$ E, then $\vdash_{\text{TNT}} \exists y(y \times 2 = \mathbf{n})$

(b)   If $n \notin$ E, then $\vdash_{\text{TNT}} \neg\exists y(y \times 2 = \mathbf{n})$.

A k-ary function f is said to be representable in **TNT** if and only if there exists a formula $\alpha[x_1, \ldots, x_k, x_{k+1}]$ in the language of **TNT**, with exactly $k + 1$ free variables $x_1, \ldots, x_k, x_{k+1}$, such that, for each $k+1$-tuple of numbers $<n_1, \ldots, n_k, n_{k+1}>$, if $f(n_1, \ldots, n_k) = n_{k+1}$, then

(a) $\vdash_{\text{TNT}} \alpha[x_1/\mathbf{n}_1, \ldots, x_k/\mathbf{n}_k, x_{k+1} / \mathbf{n}_{k+1}]$

(b) $\vdash_{\text{TNT}} \forall x_1, \ldots, x_k \exists! x_{k+1} \alpha[x_1, \ldots, x_k, x_{k+1}]$.

Clause (b) represents the fact that any function, by definition, never assigns more than one value for any given argument(s). The numerically defined quantifier "$\exists! x$" is to be read "there exists at most one x, such that ..." and can be defined via the standard existential quantifier and the identity sign. It is easy to show that an n-ary relation R is representable in **TNT** if and only if its characteristic function c_R is. Consequently, in addition, when formal representability is at issue one can talk only in terms of functions, or only in terms of relations.

One can easily show that addition and multiplication are operations represented in **TNT** precisely by the symbols which would be expected. For instance, it can be shown that, given three numbers m, n, and p, if $m + n = p$, then

(a) $\vdash_{\text{TNT}} \mathbf{m} + \mathbf{n} = \mathbf{p}$

(b) $\vdash_{\text{TNT}} \forall x \forall y \exists! z(x + y = z)$.

As I have introduced the symbols of the **TNT** language, I have spoken of the intuitive interpretation of + as the sum, etc., etc. As so remarked, the symbols and axioms of **TNT** are not randomly chosen: we want "+" to stand precisely for addition, "=" for equality, "×" for multiplication, and so on. Once we have a precise characterization of the representation of notions of intuitive arithmetic within our **TNT**, and after having established the relevant facts, we are allowed to claim that + represents addition, that = represents equality, etc., etc. Instead of relying on an intuitive interpretation, we take into account the way those symbols behave within the formalism or, more precisely, in the arrangement of the theorems of **TNT** – which in its turn depends on the **TNT** axioms and rules of inference. This is what allows us to claim that "+" does the job a good addition symbol is expected to do, that "=" does the job of the identity symbol, etc.

3   … and the Gödelian loop

Once all these (rather tiresome) formal points have been made, one can fully appreciate the importance of what Gödel achieved by means of his paper's Theorem V. This Theorem (sometimes called the Gödel Lemma) can be re-examined here by saying that *all primitive functions-relations are representable in* **TNT**: all primitive recursive truths have a corresponding **TNT** theorem, and all primitive recursive falsities have a negation which is a corresponding **TNT** theorem. When a formal system has such ability, logicians usually claim that it is *sufficiently strong*. Our Typographical Number Theory, then, is sufficiently strong. (For what? Strong enough to fall under Gödelian incompleteness, therefore showing that what counts as strength from a certain viewpoint may be inescapable weakness from a certain other.)

Today we know that, in fact, one can achieve a stronger result: **TNT** can represent all the recursive functions (that is, also the general functions). But to prove Gödel's Theorem with respect to **TNT** one only needs the representability of the primitive functions (which is why, as has been claimed in the previous chapter, we are less interested in the general recursive functions). In any case, showing that this is the case takes some time, and I will limit myself to referring you to textbooks of advanced logic for the formal details of the proof.[5] In fact, Gödel himself provided only the draft of a proof of his Theorem V in his paper. However, the result is significant precisely

---

[5]   Such as Boolos, Burgess, and Jeffrey (2002), Ch. 16.

because it completes our *loop* around the Typographical Number Theory, in the following sense.

(a)   On the one hand, two chapters ago we saw how, by means of the arithmetization of syntax, one can turn discourses on the syntax of formal systems (and therefore discourses on **TNT** too) into discourses on natural numbers. Specifically, some syntactic notions concerning the expressions of **TNT**, such as those of formula and proof, can be associated with decidable (and therefore, given Church's Thesis, recursive) arithmetical notions. Recursive functions-relations allow us to encode the syntax of the formal system.

(b)   On the other hand, Theorem V or the Gödel Lemma now guarantees that recursive arithmetical notions are representable within **TNT** itself in a precise sense. Our Typographical Number Theory, then, can represent significant portions of its own syntax "indirectly": they are syntactic notions that have been paired to recursive arithmetical ones, and these are formally representable (by means of numerals) in **TNT**:

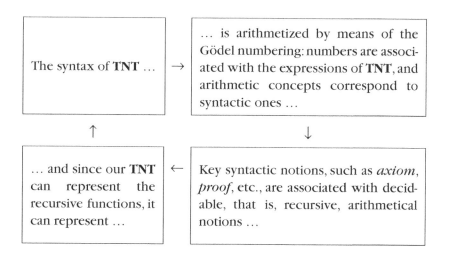

The syntax of the theory is therefore "mirrored" within the theory itself, and metamathematics is reflected within formal arithmetic. The Gödelian symphony undergoes a *crescendo*. ...

6

"I Am Not Provable"

The mathematical machinery we are now equipped with will help us understand the successive movement of the symphony. The main character of this book, that is, the Gödel sentence, is about to take the stage. It is generated by our Typographical Number Theory: a formal system composed of strings of symbols, which should not "talk about" anything or, at most, should "talk about" numbers, but should be clever enough to be capable of introspection – even of wrapping itself in a puzzle it cannot resolve. Let us look closely at the birth of this unexpected son.

1   Proof pairs

I have described how Gödel defined his list of 45 recursive items, showing the representability via arithmetization of several syntactic notions of our **TNT** in **TNT** itself. Entry 45 in the series corresponds to a two-place arithmetic relation – let us call it Prf_{TNT} – holding between two numbers, m and n, if and only if m is the Gödel number of a proof of **TNT**, and n is the Gödel number of the formula of which the proof is a proof. In other words, one can read "$Prf_{TNT}(m, n)$" as: "m is the Gödel number of a proof, in Typographical Number Theory, of the formula whose Gödel number is n." When such a claim is true of numbers m and n, I shall say that m and n form a *proof pair* of **TNT**.[1] Such a relation encodes precisely the syntactic relation that holds between a proof of **TNT** and the formula proved by it. This is why it has been labeled "Prf_{TNT}", following the standard practice of Gödelian

---

[1]   Also this appropriate terminology comes from Hofstadter's *Gödel, Escher, Bach*.

textbooks: it reminds us of the metamathematical notion encoded by the arithmetic relation. This is a (primitive) recursive relation, and therefore a decidable one; this we shall take for granted, but Gödel showed it in detail in his paper. As usual, a two-place relation is taken as a set of ordered pairs. Thus, we can likewise talk of the decidable set of *proof pairs* – let's call it D: D is the very remarkable set to which a given ordered pair <*m, n*> belongs if and only if ..., etc.[2] By the Gödel Lemma, the relation Prf_{TNT} is, then, (numeralwise) representable in our Typographical Number Theory. It will be represented by a formula with two free variables – say $\mathbf{Prf}_{TNT}(x, y)$.

Again, let us not overlook the distinction between boldface letters and italics. The expression "Prf_{TNT}" (italics) designates the two-place relation of intuitive arithmetic just introduced. The expression "$\mathbf{Prf}_{TNT}$" (boldfaced) is its formal counterpart: it is the expression representing that relation within **TNT**, in compliance with the definition of representability (by means of numerals) we encountered in the previous chapter. We have, then, that given two numbers *m* and *n*:

(a) If $< m, n > \in D$, then $\vdash_{TNT} \mathbf{Prf}_{TNT}(\mathbf{m}, \mathbf{n})$

(b) If $< m, n > \notin D$, then $\vdash_{TNT} \neg\mathbf{Prf}_{TNT}(\mathbf{m}, \mathbf{n})$.

2   The property of being a theorem of TNT (is not recursive!)

We can now define an arithmetic predicate Tb_{TNT} as follows:

$$Tb_{TNT}(n) =_{df} \exists m Prf_{TNT}(m, n).^{3}$$

Tb_{TNT} is a property of numbers, or a set of numbers. Which ones? Given the definition, we can read it as the property a given number *n* has, if and only if it is the Gödel number of a **TNT** formula for which there

---

[2]   Showing that Prf_{TNT} is a recursive relation, that is, that D is a recursive set, is not too difficult once one has established that the set of *axioms* of **TNT** is recursive. As we know, a key requisite for a real formal system is that the set of its axioms be decidable (be it a finite set or not).

[3]   Where "$=_{df}$" is read "is defined as."

exists a proof within **TNT** (less tersely: for which there exists a number *m*, such that *m* is the Gödel number of a proof in **TNT** of the formula whose Gödel number is *n*).The property at issue corresponds (via arithmetization) to that of being a theorem of **TNT**, that is, the relevant set is the set – call it T – of (the Gödel numbers of) the theorems of **TNT**.To be a theorem of our Typographical Number Theory, in fact, is just to be a **TNT** formula for which there is a proof in **TNT**.

Is this an effectively decidable, hence recursive, property? The answer is in the negative, and this is a decisive fact.The proof relation $Prf_{TNT}(m, n)$ is recursive-decidable: given any two numbers *m* and *n*, we can always effectively decide if the proof relation holds between them.We can establish whether *n* is the code of a formula and, in this case, of which; we can establish whether *m* is the code of a sequence of formulas; and by examining their components we can decide whether the latter actually is a proof of the former.

But $Th_{TNT}(n)$ has been defined as $\exists m Prf_{TNT}(m, n)$. Now this latter expression includes an unbounded existential quantification: it tells us only that *there is* (a code *m* of) a proof of the formula (whose code is) *n*. As such, it designates a property – that is, a set: the set T of (the Gödel numbers of the) theorems – which is *semi*-decidable, or recursively enumerable, but not recursive.The algorithmic test to establish whether a given formula in the language of **TNT** is a theorem is a merely partial one. We do have a procedure to systematically, "typographically" produce all the members of our set T; but we lack a decision procedure. For a given number *n*, *if* it actually is the code of a provable formula, then we can in principle show that this is the case. We can begin to mechanically, systematically produce the **TNT** proofs, and sooner or later (the code *m* of) the proof looked for will show up. But *if n* ∉ T, then our search will go on forever. In other words: when faced with an alleged proof in **TNT**, we can always decide if it actually is (which follows from the definition of formal system, and from the fact that our **TNT** is one). Given an alleged theorem τ of **TNT**, we can mechanically generate the **TNT** proofs, and check, for each of them, whether τ appears in the last line. If this happens sooner or later, then τ actually is a theorem. But one cannot foresee if and when such a proof will show up; and if τ is not a theorem, the quest will continue forever.

Prf_{TNT} is represented in **TNT** by the formula $\mathbf{Prf_{TNT}}(x, y)$; and to Th_{TNT}, which is a property of the naturals, there corresponds in the theory a formula with one free variable – let us abbreviate it with $\mathbf{Th}_{\mathbf{TNT}}(y)$ (that is: $\mathbf{Th}_{\mathbf{TNT}}(y) =_{\mathrm{df}} \exists x \mathbf{Prf}_{\mathbf{TNT}}(x, y)$).[4]

3 Arithmetizing substitution

Let us now consider an arithmetic function, $Sub(m, n, p)$, such that, if m is the Gödel number of a **TNT** formula α, and n is the Gödel number of a variable x, its value is the Gödel number of the formula $\alpha[x/\mathbf{p}]$. This is the formula one obtains from α (whose code is m), by replacing in it the free occurrences of the variable x (whose code is n) with the term $\mathbf{p}$, that is, with the numeral of number p. Just as the arithmetic relation $Prf_{\mathbf{TNT}}$ encoded a syntactic relation between formulas (more precisely: between a sequence of formulas – a proof – and the corresponding proved formula), so the arithmetic function Sub encodes a syntactic operation on formulas: the operation, familiar to anyone who knows elementary logic, of substituting the free occurrences of a variable in a given formula with a term (in the case at issue, a numeral, that is, the name of a number). This is why such a function has been dubbed "Sub": the name reminds us of the syntactic operation it encodes.

In other words, we can read "$Sub(m, n, p)$" as: "the code of the formula obtained from the formula whose code is m, by substituting in it the free occurrences of the variable whose code is n with the numeral of p."[5] And one can show that Sub is primitive recursive, and therefore,

---

[4] To be precise, $Th_{\mathbf{TNT}}$ is not properly *representable* in the technical sense defined above, but *weakly* representable. If T is our set corresponding to the property $Th_{\mathbf{TNT}}$, that is, the set of (the codes of) theorems of **TNT**, we have that:

(T) $n \in T$ if and only if $\vdash_{\mathbf{TNT}} \mathbf{Th}_{\mathbf{TNT}}(\mathbf{n})$.

Translating, this means that n is the code of a theorem of **TNT**, if and only if the formula (the sentence) $\mathbf{Th}_{\mathbf{TNT}}(\mathbf{n})$ is a theorem of **TNT**. Keep in mind that, whereas "$\mathbf{Th}_{\mathbf{TNT}}$" indicates a predicate of the language of **TNT**, "$\vdash_{\mathbf{TNT}}$" belongs to the metalanguage, meaning that what follows it is a theorem of **TNT**.

[5] I have skipped some technicalities on how to encode substitution in such a way that the substitution operation is legitimate (that is, as logicians say, in such a way as to guarantee that x is free for $\mathbf{p}$ in α).

by Gödel's Lemma, representable by an expression – call it "**Sub**" – of the formal language of **TNT**. Again, we should pay attention to the typographical scenery: "*Sub*" (italics) denotes the three-place function of intuitive arithmetic, whereas "**Sub**" (boldface) indicates the expression which formally represents the function within our Typographical Number Theory.

Here is an example that should make things clear. Suppose the formula at issue, with the free variable x and in which we want to operate the substitution, is $\exists y(y \times 2 = x)$. Let t be its Gödel number. Looking at the Gödel numbering for our **TNT** provided two chapters ago, we see that x has the number 31 as its code. Suppose we replace this variable with the numeral of *the very same* code t of the formula at issue. The result will be the formula $\exists y(y \times 2 = \mathbf{t})$ (which, by the way, "claims" that t is an even number). This formula has its own Gödel number in its turn – say v. Once one has been given the numbers t and 31, v can be effectively computed: it is the value of the recursive function *Sub*, given those numbers as arguments: $v = Sub\ (t, 31, t)$. Therefore, we can uniquely describe the number v precisely as a function of t and 31 – as the Gödel number of the formula obtained from the formula whose Gödel number is t, by substituting in it the free occurrences of the variable whose Gödel number is 31 with the numeral of the Gödel number of t itself.

4   How can a TNT sentence refer to itself?

You may wonder why I have introduced this strange recursive function *Sub*, together with the expression **Sub** that represents it in **TNT**. Accounting for this requires a small digression. You may nevertheless suspect what these formal maneuvers aim at: building within our Typographical Number Theory a self-referential sentence corresponding to the sentence G_S we met at the end of the second chapter, during our early informal exposition of the Incompleteness Theorem. Such a sentence, recall, is expected to somehow "talk of itself," in order to state its own unprovability.

In that chapter, it was claimed that the technical hitches introduced in the proof of Gödel's Theorem are largely due to the following difficulty. Self-referential sentences are easy to build in our ordinary language (in which, it seems, we can talk more or less about anything we want), for instance, by using the indexical phrase "this sentence," or analogous expressions ("I," for instance) whose reference is fixed contextually. Producing a non-contextual self-referential sentence within an artificial formal language expressing arithmetic is much more complicated. After all, such a language has been set up in order to talk about numbers, not about linguistic expressions. How could one of its formulas refer to itself? Now the arithmetization of the syntax of **TNT** opens up the possibility of building self-referential **TNT** sentences, in the sense of sentences that talk *about their own Gödel number* – about the code assigned them within the Gödel numbering procedure.

Then again, any self-referential sentence, even when formulated in our ordinary language, cannot achieve self-reference by explicitly mentioning itself (say by including all the linguistic symbols the sentence itself is composed of, in the same order and within quotation marks), for it would have to be longer than itself! If we attempt to quote such a sentence within itself in our ordinary language, we obtain such things as:

> (S) The sentence "The sentence consists of six words" consists of
> six words,

and (S) manifestly does not refer to itself. (S) refers to the phrase "The sentence consists of six words" and, by claiming that this latter is composed of six words, (S) is true: "The sentence consists of six words" is actually composed of six words. But (S) is not composed of six words.[6] (S) does not refer to itself: (S) makes a true claim on a sentence different from itself.

Now, a somewhat analogous situation takes place in a formal language. A **TNT** sentence can have self-referential features in so far as it refers to its own Gödel number; and we know that numerals play the function of number names within **TNT**. But the Gödel sentence we are searching for, and which corresponds to the G_S of my initial informal

---

[6] Of how many, then? This depends on what one means by "word" – and may also depend on the solution of sophisticated problems concerning quotation and mentioning – but we might claim that (S) is made at the very least of seven words, if we take (the token of) "The sentence consists of six words" (appearing in (S), quotation marks included, as the name of the mentioned sentence) roughly as an atomic, simple expression.

presentation of the Incompleteness Theorem, cannot refer to itself by using its own numeral (that is, by directly employing the "official" name of its Gödel number), for it would simply not fit in: no **TNT** string can include the numeral of its own code, for that numeral contains more symbols than the string itself.[7]

How do we work this out? The sentence we are looking for should refer to itself indirectly. This can be achieved via some kind of definite description: a description which, given the formal construction at issue, is to be satisfied precisely by the Gödel sentence itself. What we need is a sentence – say G – which makes the following claim:

> (G) The sentence obtained by such-and-such substitution operation has the property of being unprovable,

where, once the substitution at issue has been performed, it turns out that the sentence in question is precisely G itself. Such a sentence, thus, "talks about itself" in the sense of claiming that a sentence uniquely satisfying a certain description (that is, something like "the sentence obtained ...") has the property of being unprovable; and specifically G is the unique sentence satisfying the description.

In a note to the first paragraph of his 1931 paper (therefore, in a context in which he was still providing an informal exposition of the situation), Gödel described the fact, with some irony, thus:

> Contrary to appearances, such a proposition involves no faulty circularity, for initially it [only] asserts that a certain well-defined formula (namely, the one obtained ... by a certain substitution) is unprovable. Only subsequently (and so to speak by chance) does it turn out that this formula is precisely the one by which the proposition itself was expressed.[8]

This can be achieved in a precise way within **TNT** via our expression **Sub**. This formally represents in **TNT** a recursive arithmetic operation, that is, *Sub*. But such operation in its turn is precisely the numerical encoding, obtained via arithmetization, of a linguistic operation on the expressions of **TNT**: the syntactic operation of substituting, in a

---

[7] Intuitively, for instance, the code of such a string as, say, "The sentence with code n is not a theorem of **TNT**," would not be n, for it would be bigger.

[8] Gödel (1931), p. 42.

given formula, the free occurrences of a variable with a term of a certain kind; an operation, it has been said, familiar to anyone who has attended a course of basic logic and knows the first-order formal languages. By showing that such an operation can be arithmetically encoded, Gödel obtained a version of *G* in the formal language he was working on.

It is time for us to do the same: let the Gödel sentence for **TNT** finally take the stage!

5 γ

Consider the following formula of our Typographical Number Theory – I shall label it "γ(*y*)", for it contains the free variable *y*:

$$(\gamma(y)) \quad \neg \exists x \mathbf{Prf}_{\mathrm{TNT}}(x, \mathbf{Sub}(y, 33, y)).$$

If you read γ(*y*) in ordinary language, you get something like: "There's no *x* such that *x* is the code of a proof in **TNT** of the formula (whose code is the code of the formula) obtained from the formula with code *y*, by replacing in it the free occurrences of the variable with code 33 with the numeral of its own code."

Meandering? According to the Gödel numbering for **TNT** provided two chapters ago, the variable which was assigned the code 33 is indeed *y*. Therefore, **33** is (the short form for) the numeral of the Gödel number of that variable.[9] We can read γ(*y*), thus, as claiming that a certain formula, obtained via the substitution at issue, is not provable within **TNT**. Notice that this does not pick out *one* specific formula yet, for *y* occurs free in γ(*y*). As such (and as usual with the free variables of formal languages), *y* works a bit like a pronoun, that is, it does not have a fixed denotation. What γ(*y*) does have, just like any other **TNT** expression, is its own beloved Gödel number. It has to be a huge number, but we don't care about that – we are just content to know it is there. Let us call it *q*, and take **q** as (the short form for) its numeral.

---

[9] As you should recall, the official numeral for 33 would be 0'''''''''''''''''''''''''''''''''' (with 33 apostrophes).

Next step: substitute in $\gamma(y)$, for its free variable y, precisely the numeral **q**. We get the following formula (a sentence: no more free variables around), which we shall call γ:

(γ) $\neg\exists x \mathbf{Prf}_{\text{TNT}}(x, \mathbf{Sub}(\mathbf{q}, 33, \mathbf{q}))$.

Bingo. How should we read γ in ordinary English? It goes like this: "There's no x such that x is the code of a proof in **TNT** of the formula (whose code is the code of the formula) obtained from the formula with code q by replacing in it the free occurrences of the variable with code 33 with the numeral of its own code." Still meandering? More briefly, it says: "The formula obtained from the formula with code q by replacing its free variable with code 33 with the numeral of its own code is not provable in **TNT**." Which formula is being talked about?

Precisely γ. As it happens, $\gamma = \gamma(y)[y/\mathbf{q}]$, that is to say, γ just *is* the formula one obtains from $\gamma(y)$ (that is, from the formula with code q) by replacing its free variable y (which is the variable with code 33) with the numeral **q** (that is, with the numeral of the code of $\gamma(y)$). To put it otherwise, we can read γ as the formula which attributes the property of being unprovable in **TNT** to the formula obtained from $\gamma(y)$ by carrying out the relevant substitution operation; and γ just is the formula obtained from $\gamma(y)$ by carrying out that substitution.

γ is the Gödel sentence for **TNT** we were looking for. It is the formal counterpart of the G_S introduced in our first informal exposition of Gödel's Theorem in Chapter 2, that is, of the sentence asserting its own unprovability. One often hears that "in a word, the ... formula [γ] can be construed as asserting of itself that it is not a theorem."[10] Let us stress that "in a word." Less tersely, the situation can be described as follows: γ is a **TNT** formula, representing within the formalism an arithmetical statement. This statement arithmetizes (by way of the Gödel numbering) a syntactical or metamathematical claim on γ itself, that is: "The γ formula is not a theorem of **TNT**." It is because γ formally represents within **TNT** an arithmetical phrase, translating via the Gödel numbering a metamathematical statement which, in its turn, talks about γ, that γ is said to "talk about itself," and to claim to be unprovable in **TNT**.[11]

---

[10] Thus Nagel and Newman (1958), p. 98.

[11] See Franzén (2005), pp. 44–5 on why, opinions to the contrary notwithstanding, that γ is self-referential is completely innocent a claim. This was explicitly maintained by Gödel as well, as we shall see soon.

6 Fixed point

The strategy for constructing γ is quite close to the procedure carried out by Gödel in his 1931 paper.[12] It is a special case of a more general technique, which presupposes the arithmetization of syntax, and which has been widely used in logic since Gödel. This is usually called *diagonalization* (for it resembles Cantor's diagonal procedure, which we met in Chapter 1), or *fixed point* technique. It is precisely this method that allows us to build sentences of the kind γ, usually labeled "self-referential," within formal languages and systems. Its core lies in the following Theorem, usually called the *Diagonal* or *Fixed Point Lemma*, or the *Self-Referential Lemma*,[13] which is formulated here with reference to **TNT**. The Diagonal Lemma says that, if α[x] is any formula in the language of **TNT** with the sole variable *x* free, then there exists a sentence β such that

$$\vdash_{\text{TNT}} \beta \leftrightarrow \alpha \left[x / \ulcorner \beta \urcorner \right]$$

where, in general, given a formula α, "⌜α⌝" indicates the numeral of the Gödel number of α.[14] To put it otherwise: there exists a sentence β such that it is provable within our Typographical Number Theory that β is equivalent to the sentence obtained from α[x] by replacing its free variable *x* with the name (the numeral of the Gödel number) of β itself. β is then called a *fixed point* of α[x].[15] If α[x], as a formula with one free variable, stands in **TNT** for a set of numbers or for an arithmetic property,[16] then the Diagonal Lemma guarantees that we can prove in

---

[12] Actually, I have been sticking to the new edition of Nagel and Newman (1958), adapting it to my notation and my Gödel numbering. This new edition of the book, revised by Hofstadter, underlines the distinction between intuitive arithmetic and its formal counterpart (which I have stressed by resorting to the boldface/italics trick), thereby avoiding a certain ambiguity in the notation of the first edition.

[13] In Kleene (1986), p. 134.

[14] See e.g. Boolos, Burgess, and Jeffrey (2002), p. 221. Sometimes in the literature ⌜α⌝ is used to directly indicate the code of α, not the numeral of the code.

[15] See e.g. Smullyan (1992), pp. 102–4.

[16] One can think of every **TNT** formula α[x] as intuitively associated with a property of numbers. However, it is not guaranteed that any such formula formally *represents* a property of numbers, in the sense of "representation" defined above. This would be the case only if **TNT** were syntactically complete, but this is ruled out precisely by the Incompleteness Theorem, as we shall see very soon.

TNT that β holds iff its own Gödel number has the property α[*x*] stands for. Gödel formulated the Diagonal Lemma informally in "On Undecidable Propositions of Formal Mathematical Systems" – an important 1934 paper[17] in which he re-examined his 1931 accomplishment. In that context, he explicitly asserted the legitimacy of the self-referential reading of the relevant formulas. In the section called "Relation of the foregoing arguments to the paradoxes," he claimed:

> We have seen that in a formal system we can construct statements about the formal system, of which some can be proved and some cannot, according to what they say about the system. We shall compare this fact with the famous Epimenides paradox ("Der Lügner") ... The solution suggested by Russell and Whitehead [sc. in *Principia mathematica*], that a proposition cannot say something about itself, is too drastic. We saw that we can construct propositions which make statements about themselves, and, in fact, these are arithmetic propositions which involve only recursively defined functions, and therefore are undoubtedly meaningful statements. It is even possible, for any metamathematical property *f* which can be expressed in the system, to construct a proposition which says of itself that it has this property.[18]

Proving the Diagonal Lemma in its general form is far from obvious; but γ clearly is just a fixed point of the (un-)provability predicate. Formally, we have:

$$\vdash_{TNT} \gamma \leftrightarrow \neg \exists x \, \mathbf{Prf}_{TNT} \left(x, \ulcorner \gamma \urcorner \right)$$

or, equivalently:

$$\vdash_{TNT} \gamma \leftrightarrow \neg \mathbf{Th}_{TNT} \left(\ulcorner \gamma \urcorner \right).$$

Translating into ordinary English, this means that it is provable within **TNT** that γ is equivalent to "γ is not provable," or to "γ is not a theorem."[19]

---

[17] Actually, it consists in a series of lectures given at the Institute for Advanced Study, which were recorded, revised, and then published in Davis (1965).

[18] Gödel (1986), p. 362.

[19] The Diagonal Lemma has subsequently had an impressive variety of applications. For instance, a few years ago it was proved, by means of an argument based upon

7   Consistency and omega-consistency

At the end of Chapter 2 it was stated that, when proving in all detail
what he had drafted in the initial section of his paper, Gödel replaced
the semantic notion of soundness or correctness for the formal system
at issue (that is, its proving only truths) with a weaker assumption –
specifically, one of a syntactic kind. As you may recall, many in those
days found the concept of truth suspect, and a proof explicitly involving it would have been rejected by those who assumed the formalist
viewpoint of Hilbert's metamathematics. So Gödel replaced soundness
with a kind of arithmetic consistency, called *omega-consistency*. Let us
have a closer look at this new notion.

We already know what the claim that a formal system S is (simply)
inconsistent means: S proves both a formula of its formal language L,
and its negation. A system S is said to be *omega-inconsistent* if and only
if, for some formula $\alpha[x]$ of its language, $\vdash_s \exists x \alpha[x]$ but, for every natural
number n, $\vdash_s \neg\alpha[x/\mathbf{n}]$ (that is: $\vdash_s \neg\alpha[x/\mathbf{0}]$, $\vdash_s \neg\alpha[x/\mathbf{1}]$, $\vdash_s \neg\alpha[x/\mathbf{2}]$, …
etc.). Speaking in informal English: an omega-inconsistent system is
one that "claims" (proves) that there is some number for which a given
property or condition $\alpha[x]$ holds; but it "denies" (that is, it refutes, that
is, it proves the negation of the statement) that 0 is such a number, that
1 is such a number, that 2 is such a number, …, and so on. A system is
said to be *omega-consistent*, of course, if and only if it is not omega-
inconsistent.

Omega-consistency is stronger than (simple) consistency. Any inconsistent formal system is omega-inconsistent,[20] but not the converse:

---

diagonalization, that no antivirus program can be at the same time absolutely effective
(capable of detecting any program which might alter the code of the system of the
computer it runs on) and safe (that is, guaranteed not to alter the code of the system);
on this, see Dowling (1989). It is standard practice in mathematics to call x a fixed point
of a function f iff x is a point which is mapped to itself by the function, that is, $f(x) = x$.
And we have fixed point theorems – that is, theorems that prove that some function has
some fixed point – embedded and employed, for example, in analysis, discrete mathematics, and computer science (from Banach's Fixed Point Theorem to Kleene's
Recursion Theorem, which is a basic result in recursion theory).

[20]   See Gödel (1931), p. 30. The simplest way to show this is via the remark that, if a
formal system is inconsistent, that is, it proves both a formula and its negation, then it can
prove anything (in particular, omega-inconsistencies). This is due to a famous logical

a system can be omega-inconsistent, but consistent (we shall understand later why this is so, and why systems of this kind can turn out to be quite interesting for logicians).

8   Proving G1

We are now close to the peak: the First Incompleteness Theorem, G1 (corresponding to Gödel's Theorem VI), is in sight. Let us begin by announcing the theme; we will then perform that movement of the Gödelian symphony consisting of its demonstration. The First Theorem is the conjunction of the two following claims:

(G1a)   If **TNT** is consistent, then γ is not provable in it: $\nvdash_{\text{TNT}} \gamma$.
(G1b)   If **TNT** is omega-consistent, then $\neg\gamma$ is not provable in it: $\nvdash_{\text{TNT}} \neg\gamma$.

As for the proof of (G1a): assume that $\vdash_{\text{TNT}} \gamma$, that is, assume γ to be provable in Typographical Number Theory. Then there would be a proof of γ in **TNT**. Such a proof would have, of course, its own Gödel number: call it k. But also γ has *its* own beloved Gödel number: call it g (so $g = Sub(q, 33, q)$). Therefore, the arithmetic relation Prf_{TNT} would hold between the numbers k and g, i.e., to resort to the terminology adopted above, the ordered pair $<k, g>$ would belong to the set D of **TNT** proof pairs.

---

law, usually called *Scotus' Law*, or *Pseudo-Scotus' Law* (PS). This holds in classical (and intuitionistic) logic; it is therefore embodied, as all classical logical principles, in such formal systems as **TNT**. The law claims, quite simply, that everything follows from a contradiction. As a rule of inference, it can be formulated thus:

$$\frac{\alpha, \neg\alpha}{\beta} \text{ (PS)}$$

I believe (see Berto (2006a), (2007b)) that (PS) is a *bad* logical principle, and that classical logic is wrong, at least in this respect. A logic in which (PS) fails is called *paraconsistent*, and paraconsistent logical theories have many nice and interesting applications. Of course, in this book the virtues and vices of classical logic are not at issue; however, in the final chapter I will provide some hints on the magical world of paraconsistency.

But *Prf*$_{TNT}$ is a recursive relation, and therefore one that is representable in **TNT**: it is, in fact, represented by the two-place predicate **Prf**$_{TNT}$. This means in particular, as we already know, that if $<k, g>$ $\in$ D, then $\vdash_{TNT}$ **Prf**$_{TNT}$(**k**, **g**). Now **g** is the numeral of the Gödel number of γ, that is, **g** = $\lceil$ γ $\rceil$. Therefore, $\vdash_{TNT}$ **Prf**$_{TNT}$(**k**, $\lceil$ γ $\rceil$), so by elementary logic (by existential generalization) $\vdash_{TNT}$ $\exists x$**Prf**$_{TNT}$(x, $\lceil$ γ $\rceil$). But γ is equivalent to the negation of the latter formula, that is, to $\neg \exists x$**Prf**$_{TNT}$(x, $\lceil$ γ $\rceil$). Consequently, if we could prove γ within our Typographical Number Theory, then we could also prove its contradictory: we would have a proof both of a formula and of its negation – which is ruled out by the assumption embodied in (G1a) to the effect that **TNT** is consistent.

Now for the proof of (G1b). Assume that $\vdash_{TNT}$ $\neg$γ, that is, assume the negation of γ to be provable in **TNT**. Given that γ is equivalent to $\neg \exists x$**Prf**$_{TNT}$ (x, $\lceil$ γ $\rceil$), to say that $\neg$γ is provable amounts to claiming that $\neg\neg\exists x$**Prf**$_{TNT}$ (x, $\lceil$ γ $\rceil$) is provable, and therefore that $\exists x$**Prf**$_{TNT}$(x, $\lceil$ γ $\rceil$) is. But, as we have just established, by (G1a) γ is not provable in **TNT**, if **TNT** is consistent (and if **TNT** is omega-consistent, then it is definitely consistent, since the former property entails the latter). Therefore, for any natural number *n*, *n* is not the code of a proof of γ in **TNT** or, equivalently, for any natural number *n*, *n* does not form a proof pair with *g*: for any *n*, $<n, g>$ $\notin$ D. And again, *Prf*$_{TNT}$ is a recursive relation, and therefore one that is representable in **TNT** (the set D of proof pairs is recursive, and therefore representable in **TNT**). So from the fact that for any *n*, $<n, g>$ $\notin$ D, it follows that, for any natural number *n*, $\vdash_{TNT}$ $\neg$**Prf**$_{TNT}$ (**n**, $\lceil$ γ $\rceil$), that is to say: $\vdash_{TNT}$ $\neg$**Prf**$_{TNT}$(**0**, $\lceil$ γ $\rceil$), and $\vdash_{TNT}$ $\neg$**Prf**$_{TNT}$(**1**, $\lceil$ γ $\rceil$), and $\vdash_{TNT}$ $\neg$**Prf**$_{TNT}$(**2**, $\lceil$ γ $\rceil$), ..., and so on.

Now if **TNT** proved all these things, and also proved $\exists x$**Prf**$_{TNT}$(x, $\lceil$ γ $\rceil$), i.e., $\neg$γ, then **TNT** would be omega-inconsistent: it would prove, that is, a whole (infinite) set of sentences of the form $\neg\alpha[x/\mathbf{n}]$, and also one of the form $\exists x\alpha[x]$. Speaking informally: **TNT** would "assert" that there exists a number which is the code of a proof of γ, but at the same time, **TNT** would "deny" that 0 is that number, that 1 is that number, that 2 is that number, ..., etc. If, on the other hand, **TNT** is omega-consistent, as expressed in the assumption embodied in (G1b), then $\neg$γ is unprovable in it.

When all's said and done, if **TNT** is omega-consistent (therefore consistent), γ is an undecidable sentence of our Typographical Number Theory: **TNT** can prove neither the sentence nor its negation. Recall that, when a formal system proves neither a formula of its underlying formal language nor its negation, that is, it includes an undecidable sentence, the system is said to be syntactically incomplete. Therefore, if **TNT** is (omega-) consistent, it is syntactically incomplete.

9   Rosser's proof

Omega-consistency is a stronger assumption than (simple) consistency. In 1936, Barkley Rosser produced a variation on the First Incompleteness Theorem that does without omega-consistency. The variation consists in building a **TNT** formula which turns out to be undecidable in **TNT** (thus showing the syntactic incompleteness of **TNT**) on the basis of the sole assumption that **TNT** is (simply) consistent. Rosser's result is therefore a strengthening of Gödel's. We aren't particularly interested in the technical details of Rosser's Theorem, but some hint is appropriate. The formula at issue, which we may call the *Rosser sentence*, is slightly more complex than the Gödel sentence, and its construction exploits the well-ordered nature of natural numbers.

Generally speaking, a Rosser sentence for a suitable formal system **S**, R_S, intuitively is something like this:

(R_S) R_S is not provable in **S** before its negation, $\neg R_S$.

After the syntax of **S** has been arithmetized, each proof in **S** has its Gödel number. So we can always order the proofs in a list corresponding, say, to the order of increasing magnitude of the respective codes (given any two proofs of **S**, one comes "before" the other in the list just if the code of the former is smaller than the code of the latter). Now, assume R_S to be provable in **S**. Then there would be a proof – call it p – of R_S in **S**. If **S** is (simply) consistent, then no proof p_1 coming before p in the ordering of the proofs can be a proof of $\neg R_S$. All in all, therefore, p would constitute, together with all the proofs coming before it, a proof of "R_S is provable in **S** before $\neg R_S$", and therefore a proof of $\neg R_S$. But this is ruled out by the assumption that **S** is consistent.

Assume, then, that $\neg R_S$ is provable in **S**. Then there would be a proof – call it p again – of $\neg R_S$ in **S**. Then again, if **S** is (simply) consistent, no proof p_1 coming before p in the ordering can be a proof of R_S. Now p would constitute, together with all the proofs preceding it, a proof of "R_S is not provable in **S** before $\neg R_S$": a proof, thus, of R_S itself. But this is ruled out, again, by the consistency assumption. All things considered, R_s is undecidable on the basis of the assumption that the system is simply consistent.

The specific Rosser sentence for our Typographical Number Theory – call it ρ – is a fixed point of a more complex predicate than the **TNT** (un-) provability predicate, that is:

$$\vdash_{\text{TNT}} \rho \leftrightarrow \forall x (\mathbf{Prf}_{\text{TNT}}(x, \lceil \rho \rceil) \rightarrow \exists z (z < x \wedge \mathbf{Prf}_{\text{TNT}}(z \lceil \neg \rho \rceil))).$$

As we know, γ is equivalent to: "For all x, x is not (the code of) a proof of γ." Instead, ρ is equivalent to: "For all x, if x is (the code of) a proof of ρ, then there is a z smaller than x (that is, coming before x in the ordering), such that z is (the code of) a proof of ¬ρ."[21]

---

[21] For a detailed proof that ρ is undecidable in **TNT** given the consistency of **TNT**, see e.g. Moriconi (2001), pp. 243–6; and Boolos, Burgess, and Jeffrey (2002), pp. 225–7 (where the proof is carried out with respect to a system weaker than **TNT** and called **Q**, of which I shall talk later).

7

The Unprovability of Consistency and the "Immediate Consequences" of G1 and G2

In the last, short section of his 1931 paper Gödel produced an Eleventh (and last) Theorem. This was later to be called the Second Incompleteness Theorem. In fact, as has already been remarked towards the end of Chapter 2, it is a corollary of the First Theorem, and equally important. But the proof of G2 provided by Gödel in his paper "consisted mostly of handwaving,"[1] that is, it was but a draft.

How come? Taken in itself, the informal idea at the core of the Second Theorem is wonderfully simple, as we are about to see. The tricky part comes from the fact that a detailed proof of G2 involves some subtle features of the syntactic notions in play. It is often claimed that the two Theorems apply to formal systems which include "a certain amount of arithmetic." However, the certain amount needed to prove the First Theorem and the certain amount needed to prove the Second are not exactly the same. Specifically, in order to obtain G2 one has to show the provability within the formal system of certain internal features of its proof relation. But let's proceed step by step.

1   G2

The first half of the First Theorem, G1, has told us that

(G1a)   If **TNT** is consistent, then γ is not provable in it.

Can (a formal counterpart of) this very claim be proved within our Typographical Number Theory? It seems so. All in all, the argumentation

---

[1]   Franzén (2005), p. 48.

used to establish (G1a) earlier is based upon some reasoning at the level of elementary arithmetic, that is, precisely upon the kind of reasoning **TNT** is supposed to formalize. It is to be expected, then, that one can give a proof of the formula that formally represents the entailment of (G1a) within **TNT**.

But how are we to build such a formula? To this end, we need a string representing the antecedent of the conditional that (G1a) consists in; we need a "declaration of consistency" representing (by way of arithmetization) the metamathematical sentence: "Typographical Number Theory is consistent." Call such a formula **Cons**$_{TNT}$. It can be defined thus:

$$\textbf{Cons}_{TNT} =_{df} \neg\exists x \textbf{Prf}_{TNT}(x, \lceil 1 = 0 \rceil).$$

So **Cons**$_{TNT}$ is also equivalent to $\neg\textbf{Th}_{TNT}(\lceil 1 = 0 \rceil)$. Let's christen **Cons**$_{TNT}$ the *consistency statement* for **TNT**. The explanation of why it deserves such a name goes as follows.

Of course, the formula $1 \neq 0$ (that is: $\neg (1 = 0)$) is provable in our Typographical Number Theory: it is but an instance of its first axiom (TNT1), that is, $\forall x(x' \neq 0)$. So if **TNT** proved $1 = 0$ too, **TNT** would be inconsistent, proving both a formula and its negation. Conversely, if **TNT** is inconsistent, it proves $1 = 0$, because an inconsistent formal system, as we know, proves anything because of Scotus' Law.[2] All in all, **TNT** is consistent iff $1 = 0$ cannot be proved in it. "'$1 = 0$' is not provable in **TNT**" is precisely the metamathematical statement represented (via arithmetization) within **TNT** by **Cons**$_{TNT}$, that is, $\neg\exists x \textbf{Prf}_{TNT}(x, \lceil 1 = 0 \rceil)$, that is, $\neg\textbf{Th}_{TNT}(\lceil 1 = 0 \rceil)$.

Representing (G1a), that is, the claim "If **TNT** is consistent, then γ is not provable in it," is now utterly simple:

(C) $\textbf{Cons}_{TNT} \rightarrow \neg\exists x \textbf{Prf}_{TNT}(x, \lceil \gamma \rceil)$.

---

[2] For this reason Gödel in his paper proposed as a consistency statement for the system he was working on a formula formally representing the claim "Some formula is unprovable in the system" (see Gödel (1931), p. 45). Such a claim is equivalent to a consistency statement precisely because, were the system inconsistent, it would prove anything by means of Scotus' Law. For our Typographical Number Theory this would correspond to something like $\exists y \neg\exists x \textbf{Prf}_{TNT}(x, y)$, that is, $\exists y \neg\textbf{Th}_{TNT}(y)$ (this is the path followed, for instance, by Nagel and Newman (1958), pp. 104–5). Sometimes some other inconsistent sentence is used instead of $1 = 0$ (see e.g. Smullyan (1992), p. 108; Moriconi (2001), p. 221). These differences are of minor importance, though.

But the consequent of this conditional is equivalent to … γ! Therefore, (C) just is: $\textbf{Cons}_{\text{TNT}} \to \gamma$. Now we quickly obtain Gödel's Second Incompleteness Theorem, which goes thus:

> (G2)   If **TNT** is consistent, then $\textbf{Cons}_{\text{TNT}}$ is not provable in it: $\nvdash_{\text{TNT}} \textbf{Cons}_{\text{TNT}}$.

That is to say: if **TNT** is consistent, then it cannot prove the formula, $\textbf{Cons}_{\text{TNT}}$, formally representing in **TNT** itself the claim that **TNT** is consistent. Even more concisely: **TNT** cannot prove its own consistency.

The key step in a detailed proof of the Second Theorem consists in proving that (C) is itself a theorem *of* **TNT**: not only does **TNT** fall under (the first half of) the First Theorem, (G1a), but also, so to speak, it "is aware" of this fact. Things being so, if we could prove the antecedent of that conditional, that is, $\textbf{Cons}_{\text{TNT}}$, then we could also run a simple *modus ponens* and prove the consequent, γ. We would have, that is, a simple proof like the following:

> (1) $\textbf{Cons}_{\text{TNT}} \to \gamma$ **TNT** theorem
> (2) $\textbf{Cons}_{\text{TNT}}$ **TNT** theorem?
> (3) γ From (1) and (2), by *modus ponens*.

However, that such a proof is available has been ruled out by (G1a), that is, by the first half of the First Theorem.

The counterpart for our Typographical Number Theory of what Gödel didn't do, but conjectured to be feasible with respect to the system he was working on,[3] consists in proving precisely that the proof of (G1a) can be formalized within **TNT**, thereby obtaining $\textbf{Cons}_{\text{TNT}} \to \gamma$ as a theorem of the system. The title of Gödel's 1931 paper, in fact, ends with the Roman number "I" because Gödel was planning to write something like "On Formally Undecidable Propositions II," including in it a full-fledged proof of G2. But the paper was never written – partly because of the fast acceptance of G1 and G2 by scholars,[4] and partly because the result was

---

[3]   See Gödel (1931), pp. 39–40.
[4]   People had a hard time *understanding* the technical details and sometimes even the overall sense of G1 and G2 – mainly because, as I have said, Gödel's techniques were quite innovative for the time. But once they had been understood, G1 and G2 were usually accepted (by Hilbert as well) with no further discussion. There were a few exceptions,

obtained by Hilbert and Bernays a few years later.[5] The fact that a completion of the proof of G2 had the signature of Hilbert himself was an important confirmation of Gödel's result.

In **TNT** it is possible to prove, not only that $\mathbf{Cons_{TNT}} \to \gamma$, but also the converse entailment, $\gamma \to \mathbf{Cons_{TNT}}$.[6] This is important because it tells us that $\vdash_{TNT} \mathbf{Cons_{TNT}} \leftrightarrow \gamma$, that is, one can prove in **TNT** that the Gödel sentence for **TNT** is equivalent to its consistency statement $\mathbf{Cons_{TNT}}$ (and, of course, just as γ is an undecidable sentence of **TNT**, $\mathbf{Cons_{TNT}}$ is too).

2 Technical interlude

Now comes a technical detour, of minor importance in the context of our main trip. It has been found that, in fact, one doesn't need to formalize the whole proof of (G1a) in order to get a proof of the Second Theorem. One "only" needs to formalize some auxiliary statements expressing conditions on the encoding of the provability predicate $\mathbf{Th_{TNT}}$. These are the following: for any two given formulas α and β,

(P1) If $\vdash_{TNT} \alpha$, then $\vdash_{TNT} \mathbf{Th}_{TNT}(\lceil \alpha \rceil)$

(P2) $\vdash_{TNT} \mathbf{Th}_{TNT}(\lceil \alpha \rceil) \to (\mathbf{Th}_{TNT}(\lceil \mathbf{Th}_{TNT}(\lceil \alpha \rceil) \rceil))$

(P3) $\vdash_{TNT} \mathbf{Th}_{TNT}(\lceil \alpha \to \beta \rceil) \to (\mathbf{Th}_{TNT}(\lceil \alpha \rceil) \to \mathbf{Th}_{TNT}(\lceil \beta \rceil))$.

Nowadays, logicians tend to call a "provability predicate" in the strict sense a predicate satisfying such conditions as (P1)–(P3).[7]

---

of which I shall talk later. There were also some initial qualms about the paternity of the Incompleteness Theorem, which a guy called Paul Finsler claimed to have proved before Gödel. We aren't interested in this issue, since nowadays Finsler's result is not even considered a minor anticipation of Gödel's.

[5] In Hilbert and Bernays (1939), pp. 283–340.

[6] Informally: γ "asserts" that γ is not provable, and the claim that some formula is unprovable entails the claim that **TNT** is consistent, for if it weren't, then it would prove anything.

[7] Feferman (1960) is a wonderful classic work on these issues.

Translating the whole thing into ordinary English: (P1) just claims that if a given formula α is a theorem of **TNT**, then the formula $\mathbf{Th}_{TNT}(\!\lceil\alpha\rceil\!)$, formally representing this fact, is also a theorem. So far, no surprise.

As for (P2): the formula $\mathbf{Th}_{TNT}(\!\lceil\alpha\rceil\!) \rightarrow (\mathbf{Th}_{TNT}(\!\lceil\mathbf{Th}_{TNT}(\!\lceil\alpha\rceil\!)\rceil\!))$ formally represents the fact I have just mentioned, that is, that if α is a theorem, then the formal deputy of this in **TNT**, $\mathbf{Th}_{TNT}(\!\lceil\alpha\rceil\!)$, is also a theorem. (P2), therefore, claims that this is also provable in **TNT**. (P2) actually is the difficult thing to prove, and the proof usually has no place in inter-mediate textbooks,[8] for it is painstakingly long.

Finally, (P3) just expresses the provability of the closure of our Typographical Number Theory under *modus ponens*: the formula $\mathbf{Th}_{TNT}(\!\lceil\alpha\rightarrow\beta\rceil\!) \rightarrow (\mathbf{Th}_{TNT}(\!\lceil\alpha\rceil\!) \rightarrow \mathbf{Th}_{TNT}(\!\lceil\beta\rceil\!))$ represents the fact that, if the formulas $\alpha \rightarrow \beta$ and α are theorems of **TNT**, then β is too. This is clearly the case, since **TNT** embodies *modus ponens*; and (P3) tells us that the representation of this is a theorem of **TNT**.

3   "Immediate consequences" of G1 and G2

We are now close to the end of our symphony. The journey has been a long one, the music sometimes difficult to listen to. We have been digging deep into the details of that amazing ride of abstract thought which is Gödel's Theorem. Before we conclude this first part of our book, though, something has to be said on the immediate consequences of G1 and G2.

The expression "immediate consequences" should be taken as inten-tionally vague. It means something like "*relatively* non-controversial upshots of the theorems" – things following from them quite directly, and on which mathematical logicians have a strong tendency to agree (of course, some of these upshots have been challenged by some phi-losophers of mathematics and logic; but then, philosophers dispute about anything). The whole second part of the book will be devoted to the exploration of a bunch of *less* immediate consequences of G1 and G2: the readings and interpretations proposed by philosophers, math-ematicians, physicists, and yet others on what Gödel's results would

---

[8]   See e.g. Smullyan (1992), p. 107; Boolos, Burgess, and Jeffrey (2002), pp. 233–5; but not Moriconi (2001), pp. 215–17.

entail for mathematics, philosophy, our understanding of our mind, of ourselves, and of the world in general. I shall show there that some of these alleged "consequences" do not follow at all; that some others follow only when we buy some further, disputable assumptions; and that others yet, albeit debatable, are decidedly fascinating.

At the outset, one should claim that Gödel's *First* Theorem has a certain impact on Hilbert's Program. Hilbert's famous *non ignorabimus* involved the demand that any mathematical problem be decidable, whereas G1 provides evidence of the existence of undecidable sentences within a system specifically built for arithmetic, such as **TNT**. So far, though, nothing drastic has happened. We must bear in mind that γ is not an absolute *ignorabimus* at all. In fact, no Gödel sentence for a given formal system **S** constitutes an absolute *ignorabimus* merely because it is undecidable in **S** (actually, there are some quite subtle issues here; but I'll postpone their discussion to the second part of the book). What the First Theorem shows is only that one cannot exhibit a *single* sufficiently strong formal system within which all the mathematical problems expressible in the underlying formal language can be decided.

This said, it is the Second Theorem that, as is commonly admitted, collides with the formalist program. As we have seen some chapters ago, once the heart of classical mathematics had been formalized, Hilbert wanted to prove metamathematically the consistency of the resulting system by finitary reasoning. Hilbert was never completely explicit in his characterization of "finitary methods." However, one can certainly expect these methods to be formalizable in such systems as **TNT**, given they appear to be weaker than those available in **TNT** itself. G2 rules out this possibility: a consistency proof for **TNT** requires methods of proof unavailable (that is, not formalizable) within the system. People usually say: *stronger* methods. In a sense, this claim is true; but in another sense, it calls for qualification. This will be discussed in the second part of the book, when we come to talk about the skeptical interpretations of the Second Theorem.

4   Undecidable$_1$ and undecidable$_2$

According to G1, as we have seen, **TNT** is syntactically incomplete, that is, it includes undecidable formulas. One should not confuse

(1) undecidability as a property of *formulas* with (2) undecidability as a feature of *formal systems*.

(1)   The first property, which we know well by now, is always relative to a given formal system **S**: to assert that α is undecidable in **S** is to maintain that, *in* **S**, α is not provable and not refutable, that is, such that its negation is not provable either. This does not rule out α being provable (therefore decidable) elsewhere. This point has already been flagged before, and it depends on the following simple remark: talk of formally *absolutely* undecidable formulas is sort of meaningless. Of course, mathematicians harmlessly claim to have proved, or refuted, some mathematical statement or other, feeling no need to add any qualification. What is at issue in this case, though, is what I have dubbed "intuitive" mathematics. On the other hand, given any formula α, one can always build a formal system in which that formula is provable (provable, that is, in the formal sense – in the sense of proof theory). One can always construct (or find) a formal system that has α as a theorem, at the very least because the axioms of formal systems are theorems; this is one of the cases in which the terminology of formal systems departs from that of informal mathematics, where if something is an axiom, it is not a theorem, that is, the conclusion of a proof. Therefore, building a formal system having α as an axiom is sufficient to obtain, trivially, that α is provable (therefore decidable) in that formal system, in spite of its being undecidable somewhere else.[9]

(2)   Undecidability in the second sense, on the other hand, is a property of any formal system **S** whose set of theorems is not decidable, that is, given Church's Thesis, recursive. Given a formula α in the language **L** of **S**, there is no effective procedure terminating in a finite number of steps which allows us to decide whether α is a theorem of **S** or not. Now our Typographical Number Theory is an undecidable formal system in this sense. The Fundamental Property of formal systems, recall, prescribes that the set of theorems of a real formal system

---

[9]   This is already an "immediate consequence" about which some philosophers might complain – Wittgensteinians, for instance. Wittgenstein thought it absurd to claim that the same formula α is undecidable in some formal system, but decidable in some other. When we modify a system, according to Wittgenstein, α is not the same formula anymore (that is, it cannot retain the same meaning it had in the previous system), even though it stays typographically identical. I shall talk about this peculiar Wittgensteinian stance in the final chapter of the book.

be at least *semi*-decidable, that is, recursively enumerable: its theorems must be mechanically, effectively producible. However, the arithmetic property Th_{TNT} (that is, the set of theorems of **TNT**) is only recursively enumerable. We can produce the **TNT** theorems in a mechanical way, but we cannot produce the **TNT** *non*-theorems in a mechanical way.

This said, there is an important connection between undecidability$_1$ and undecidability$_2$: a (consistent) syntactically complete formal system **S** is always decidable. For given any formula α in the language of **S**, either α or its negation is a theorem; this is what **S**'s syntactic completeness consists in. Therefore, the set of theorems of **S** is decidable: for any given formula α, one can establish whether α is a theorem or not by enumerating the theorems of **S**, and sooner or later one will stumble upon α, or upon $\neg\alpha$ (and if one stumbles upon $\neg\alpha$, then, given the consistency of the system, α is not a theorem).

5   Essential incompleteness, or the syndicate of mathematicians

One could believe the incompleteness of our Typographical Number Theory to be a contingent fact. One could think, so to speak, that the formal system limps, has not been built properly, and needs a crutch. Perhaps something is missing, maybe some axiom. In this case, things could be fixed quite easily: we may just add to **TNT** its Gödel sentence γ, precisely as an axiom! Let's consider, then, the system $\mathbf{TNT}_1 = \mathbf{TNT} + \gamma$, obtained precisely by adding γ as an axiom to our Typographical Number Theory.

Can $\mathbf{TNT}_1$ circumvent incompleteness? The answer is no, because $\mathbf{TNT}_1$ in its turn is sufficiently strong, that is, capable of representing the recursive functions-relations. Therefore, we have the conditions for reapplying to $\mathbf{TNT}_1$ the Gödelian procedure which has led to the proof of G1 for **TNT**. Of course, each time the axioms of a formal system are modified, its proof relation, that is, the set of proof pairs of the system, also changes. In the specific case, we now have a relation, say Prf_{TNT1}, a bit different from Prf_{TNT}. Starting with this new relation, we can build a new formula γ_1 different from γ, that is, not equivalent to it, and undecidable in $\mathbf{TNT}_1$ (*its* Gödel sentence).

Let's try again then. Let's add γ_1 as a new axiom to TNT_1. We obtain a theory $TNT_2 = TNT_1 + \gamma_1 = TNT + \gamma + \gamma_1$. Same trouble: TNT_2 is sufficiently strong; now we can build *its* Gödel sentence, that is, a formula γ_2, different both from γ and from γ_1, and undecidable in TNT_2. If we move on to a theory TNT_3 obtained by adding γ_2 ... and all that, we simply reproduce the problem. The procedure can be iterated as many times as we wish. And all these formulas $\gamma_1, ..., \gamma_n$ are different, in the sense of not being equivalent to each other. Since **TNT** is a subsystem of all the systems in the chain, $TNT_1, ..., TNT_n$,[10] all the formulas $\gamma_1, ..., \gamma_n$ are undecidable in our Typographical Number Theory too. As a matter of fact, **TNT** faces an infinity of non-equivalent undecidable formulas.

More generally, one can establish that any consistent way of extending **TNT**, also more sophisticated than just adding by brute force the relevant Gödel sentence as an axiom, will not lead to the most wanted completeness, provided the theory delivered by the extension is still axiomatizable – a real formal system, that is. The situation is described by saying that **TNT** is *essentially* incomplete: irremediably so, as you might say.

There's a metalogical result called the *Lindenbaum Lemma* which appears to contradict this fact. The Lemma states, roughly, that any consistent theory formulated in a first-order logical language can have a consistent and complete extension. However, such an extension in the case of **TNT** could not be said to be a real formal system anymore, for it could not be recursively axiomatized: it would not satisfy, that is, the Fundamental Property of formal systems. Let us take into account, for instance, the formal theory sometimes called the "True Arithmetic."[11] This is, quite simply, the theory having as axioms all and only the arithmetical truths. Such a theory appears to "prove" all the truths of arithmetic by definition, for it has them as axioms, and axioms are theorems. But, in fact, it cannot be axiomatized: the set of axioms (and therefore the proof relation) is not recursive-decidable. Given a formula in the relevant formal language, there is no effective procedure to decide whether

---

[10] Given two formal systems S_1 and S_2, S_2 is said to be *stronger* than S_1, or an *extension* of S_1, if and only if it proves everything S_1 proved, and also something more: the set of theorems of S_1 is a proper subset of the set of theorems of S_2 (with some provisos on the underlying formal language(s) of the systems, which can be skipped here). It is also said, then, that S_1 is a *subsystem* of S_2.

[11] See e.g. Smullyan (1992), p. 57.

it is an axiom of the theory or not. Therefore, the True Arithmetic is not a real formal system. Theories of this kind are studied and have many interesting features; but they are not formal systems in the strict sense.

In a sense, one could conjecture that mathematicians may call themselves lucky because of this (in the same sense, that is, in which dentists may call themselves lucky because nobody has yet invented the Perfect Mouthwash – that is, a mouthwash which keeps everyone's teeth perfectly healthy forever). Should we have an algorithmic test for the axioms of the True Arithmetic, we would have a decision procedure for arithmetical truth: in order to establish whether a given arithmetic sentence is true or false, it would be sufficient to resort to a mechanical and always terminating procedure. Given sufficiently powerful calculators, the open problems of arithmetic would quickly reduce, and Hilbert's dream would turn into the nightmare of unemployment for lots of working mathematicians. In this sense, the essential incompleteness of **TNT** is a good Syndicate for the Defense of Mathematicians' Rights! (In fact, there are many other subtle issues lurking here: we shall have a look at them in Chapter 11, on Gödel's Theorem and artificial intelligence).

6   Robinson Arithmetic

Systems weaker than **TNT** also fall under Gödelian incompleteness. This happens in particular to a formal system for arithmetic which has received a good deal of attention among mathematical logicians, called *Robinson Arithmetic*, and usually labeled **Q** (sometimes **R**). People often prove the First Incompleteness Theorem for **Q**, and then claim that any consistent formal system extending **Q** is incomplete.

A pedagogically useful way to present the system consists in adding to the formal language of **TNT** the symbol "<" as primitive, and then in formulating the following axioms:[12]

---

[12]   Boolos, Burgess, and Jeffrey (2002), pp. 207 ff, call this axiom system **Q** *Minimal Arithmetic,* and make a distinction between it and what they call Robinson Arithmetic, which they label **R**. **Q** is not exactly equivalent to their **R**, but this minor issue is of lesser importance for us.

(Q1) $\forall x(x' \neq 0)$
(Q2) $\forall x \forall y(x' = y' \rightarrow x = y)$
(Q3) $\forall x(x + 0 = x)$
(Q4) $\forall x \forall y(x + y' = (x + y)')$
(Q5) $\forall x(x \times 0 = 0)$
(Q6) $\forall x \forall y(x \times y' = (x \times y) + x)$
(Q7) $\forall x \neg(x < 0)$
(Q8) $\forall x \forall y(x < y' \leftrightarrow x < y \vee x = y)$
(Q9) $\forall x \forall y(x < y \vee x = y \vee y < x)$

Basically, one takes the first six axioms of **TNT** (corresponding to (Q1)–(Q6)), deletes the induction scheme, and adds (Q7)–(Q9) regulating < taken as primitive. Robinson Arithmetic is a weak system. Its main disadvantage consists precisely in its lacking the induction principle: mathematical induction is a fundamental strategy in the proof of lots of important mathematical theorems. In a sense, though, this is also what makes the system interesting. We know that **TNT** captures mathematical induction only by means of a scheme, due to the limitations of the expressive power of its first-order language. But RobinsonArithmetic is *finitely* axiomatized – and, despite being so, it is still sufficiently strong, that is, Gödel's Lemma holds for it too: the system can represent the recursive functions-relations.[13] So Gödel's First Theorem applies to **Q**, and the system is incomplete.

7   How general are Gödel's results?

G1 and G2, in the end, hold for any extension of Typographical Number Theory, provided it is still a formal system, (omega-) consistent (and, of course, sufficiently strong). To sum up, the conditions for applying Gödel's Theorem to a formalized theory, in their most general formulation, can be phrased as follows:

(1)   To begin with, the theory has to be a real formal system: it has to be axiomatizable, the set of its axioms has to be a decidable one; it has to satisfy, that is, the Fundamental Property.

[13] See ibid., p. 212.

(2) Next, the theory must be sufficiently strong, i.e., capable of representing the (primitive) recursive functions.

(3) Third, the theory has to be consistent (or omega-consistent; but to prove the "Rosserian" variant of G1 simple consistency is enough).

These conditions are fulfilled also by the current mainstream set theories: specifically, by **ZF-ZFC**. I have claimed that Gödel was always careful about the generalizability of his result (even though from the beginning he indicated the extensions I have just mentioned).[14] The reason for his cautious attitude is that, as we have abundantly seen, the notion of formal system needs to be specified in close connection with algorithmic concepts: decidability, computability, etc. We need these notions, for we claim that a formal system has a decidable set of axioms, that its set of theorems has to be enumerable, etc. If a formal theory does not fulfill these requirements, it is not an authentic formal system, and may escape incompleteness just because of this.

To specify in a precise way the range of theories to which the Incompleteness Theorem applies, thus, we need an exact definition of the notion of formal system; and in order to obtain this, we need a theory that characterizes the notion of computability (of algorithm, etc.) in a precise way. Gödel waited until such a precise characterization was available; and he believed a good one to have been provided by the British mathematician Alan Mathison Turing. In order to talk about this, we need a quick depiction of the famous *Turing machine*.

8   Bits of Turing machine

The Turing machine is not an actually existing machine, but an ideal one; however, the idea it consists in, which dates back to around 1935, has been the basis for the construction of those very real machines we nowadays call computers. Turing came from Cambridge, but he spent the academic year 1936-7 at the Institute for Advanced Study in Princeton, where von Neumann was actively promoting Gödel's results.

---

[14] See Gödel (1931), pp. 33-4.

When Turing went back to Cambridge, he was probably ruminating on the Incompleteness Theorem.[15] His ingenious idea consisted in proposing, so to speak, an operational definition of the notion of algorithm, starting with what we actually *do* when we compute something.[16] If you look at a *Homo sapiens* doing some computation, you will probably see her writing down ciphers, copying them, moving down one line, restarting, etc. She will do it, typically, by mentally following a mechanical procedure she has learned by heart at school, and she will halt after writing down the result.

A Turing machine is at once a simulation of, and a way of making precise, this kind of procedure. Typically it is conceived as a tape, thought of as infinite in length (it is mainly this that makes the Turing machine an "ideal" one: the tape corresponds to the memory capacity of a computer, which of course is always finite, albeit enormous). The tape is divided into squares, each of which is occupied by a symbol taken from a finite pre-specified set (to get intuitively closer to a computer we can imagine the only symbols around to be 1 and 0, as is the case with computers counting in base two; but this is not mandatory). The machine includes a device for scanning, erasing, and writing down symbols which moves up and down the tape, one square at a time. Specifically, the computing device can do only one of the following things at a time: scanning the symbol written on a square of the tape; erasing it and writing down another (not necessarily different) one; moving to the left or to the right (or up or down: this makes no difference) by one single square. But the machine also has a finite series of *internal states*, say $s_1, \ldots, s_n$, where s_1 is called the initial state (the one at which the machine finds itself at the beginning of a computation) and s_n is called the final state (the one at which the machine *halts*).

At any moment during the computation, the machine always finds itself exactly in one of these states; but it can pass from one state to another after carrying out an *instruction*. In fact, the machine executes computing instructions, specifiable as something of the form: "If you find yourself in state s_i, and you scan the symbol σ, then overwrite

---

[15]   See Goldstein (2005), p. 195.
[16]   And his memorable paper is Turing (1937).

the symbol σ_1, move one square to the left (to the right), and pass on to the state s_j."Any instruction, therefore, has the shape of a conditional: it tells the machine what to do, depending on its current internal state and on the scanned symbol (either 0 or 1, in our case). The machine follows a deterministic procedure (as an algorithmic one is supposed to be): the instructions specify all that has to be done, for any state the machine finds itself in, and for any symbol it happens to scan (this requirement can be weakened, but we are not interested in this further possibility in our compressed presentation). Instructions also determine the next state the machine has to be in. And they are completely consistent, in the sense that it cannot be the case that two instructions associate two distinct things to do to one and the same <status, scanned symbol> pair.

The set of instructions specifies the "program" the machine consists in. Different Turing machines are individuated by the different programs or sets of instructions, and these machines can carry out additions, multiplications, exponentiations, etc., in the specified notation (in base two, in our example). I will skip the minutiae of the theory: there are many good textbooks that provide detailed explanations.[17] What interests us is that Turing machines perform mathematical computations, and that all the functions computable by some Turing machine or other are intuitively computable. The thesis that, conversely, all the effectively computable functions are Turing-computable is called *Turing's Thesis*. Just like Church's Thesis, it is considered as something that might never be demonstrated once and for all, and this for the usual reason: effective computability is just an intuitive notion.

Now, as announced several pages ago, the characterization of computable functions in terms of Turing machines and the characterization in terms of (general) recursion theory were proved *equivalent*. Turing showed that the functions computable by a Turing machine coincide with the recursive ones: given any recursive function, there is a Turing machine computing that function for any argument(s); conversely, for any Turing machine there is a recursive function the machine computes. The two theories are, so to speak, extensionally equivalent: they individuate the very same set of functions, albeit by adopting very different explanatory concepts. Consequently, Turing's Thesis coincides

---

[17]   See e.g. Boolos, Burgess, and Jeffrey (2002), Ch. 3; Frixione and Palladino (2004).

in this sense with Church's Thesis (which is why people often talk of a unique Church-Turing Thesis).

And this is not all. As an abstract, idealized entity, a Turing machine coincides with, and is fully specified by, its set of instructions. These very instructions, then, can be written down on the tape of another Turing machine once they have been suitably *encoded*, for instance as sequences of 0s and 1s. This latter machine can scan the instructions in which the former consists and, given a certain input, it can apply such instructions. Arithmetization plays a central role again: for instance, one may pick a set of instructions in which a Turing machine M consists, and encode such a set via a specific arithmetization so as to assign it a suitable code, say m. Suppose M, once given inputs $x_1, \ldots, x_n$, produces x as the corresponding output. Then there exists a Turing machine N which, given inputs $m, x_1, \ldots, x_n$ (that is, the same inputs plus the code m of M as a further one), produces precisely x as the corresponding output. This means that the second machine can "simulate" the behavior of the first, the computations M could perform. This should be familiar from ordinary computers, in which procedures, that is programs, are encoded with the same code as the data they operate on, so that programs themselves can play the function of data, manipulated by further programs.

It can be proved that there exists a *universal Turing machine*, that is, a machine capable of simulating any computation carried out by some Turing machine or other. Therefore, a universal Turing machine can compute all the recursive functions; and a computer just is or embodies the physical realization of a universal Turing machine (albeit with a finite memory), including the different particular Turing machines as programs, corresponding to the instructions representing the machines themselves.

We shall return to Turing machines and computers in the second part of the book; but now, back to Gödel.

9   G1 and G2 in general

Today we know that precise characterizations of the notions of computability and algorithm can be provided by recursion theory, if one accepts Church's Thesis. Despite being one of the founding fathers of recursion theory with his 1931 paper, though, Gödel wasn't convinced by Church's Thesis, at least in the beginning: he was reluctant to consider even general recursion as an exhaustive characterization of computability. Only after Turing had shown the equivalence between Turing-computable and recursive functions did Gödel accept that the effectively computable functions coincide with the recursive ones.

Several papers have been written to attempt an explanation of why Gödel was suspicious of Church's Thesis, whereas he wholeheartedly accepted Turing's, which is equivalent to it.[18] Of course, this was not a mistake on Gödel's part: let's just say that he didn't have much "sixth sense" on the issue, which is strange, given the genius he was. People have also wondered why Gödel had a less central role than he could have had in the subsequent momentous developments of recursion theory, and some have ascribed this fact to his mathematical Platonism[19] (to which I shall come in Chapter 9). I would rather believe that this stemmed from the deeply philosophical nature of his intellect. His famously insightful mind gave him the ability to solidly focus on the *fundamental* problems, opening new perspectives and disclosing original viewpoints. He could leave others to develop the technical niceties. In any case, in the Note added in 1963 to his 1931 paper, which explicitly provides the awaited generalization of the Incompleteness Theorems, the spotlight is still occupied by Turing:

> In consequence of later advances, in particular of the fact that due to A.M. Turing's work a precise and unquestionably adequate definition of the general notion of formal system can now be given, a completely general version of Theorems VI and XI is now possible. That is, it can be proved that in every consistent formal system that contains a certain amount of finitary number theory there exist undecidable arithmetic propositions and that, moreover, the consistency of any such system cannot be proved in the system.[20]

Here are, then, two good general-abstract formulations of G1 and G2:

(G1)   (Gödel–Rosser) Any (omega-) consistent formal system **S** capable of representing a certain portion of elementary arithmetic

---

[18]   See e.g. Davis (1982).
[19]   See e.g. Feferman (1983).
[20]   Gödel (1931), pp. 40–1.

is syntactically incomplete: there exist arithmetical statements formulated in the language **L** of **S**, such that they are neither provable nor refutable in **S**.[21]

(G2)   (Gödel) Any consistent formal system **S** capable of representing a certain portion of elementary arithmetic cannot prove the sentence of its language **L** representing the consistency of **S**.[22]

10   Unexpected fish in the formal net

We have examined what happens if we add to Typographical Number Theory γ and its cousins $\gamma_1, \ldots, \gamma_n$. In the second part of the book, I shall say something on more sophisticated ways of strengthening **TNT** which, while not being a remedy for its essential incompleteness, nevertheless lead to interesting remarks on the philosophical outcomes of the Incompleteness Theorem. But what happens if we add to the formal system the negation of its Gödel sentence, $\neg\gamma$? Things get interesting. Since γ is undecidable within **TNT**, that is, neither γ nor $\neg\gamma$ is a theorem, adding *either* to the system produces a consistent system (provided the original **TNT** was). Call the expanded system **TNT***: so **TNT*** = **TNT** + $\neg\gamma$. Now, because of the Completeness Theorem of first-order logic, by being consistent **TNT*** deserves a *model*: call it $\mathbb{M}$.[23] This

---

[21]   Recall that (simple) consistency is sufficient for the Rosser sentence, whereas omega-consistency is required for the Gödel sentence.

[22]   Recall that the "certain portion" of arithmetic is not exactly the same for G1 and G2.

[23]   I have already hinted at the fact that the various versions of first-order predicate calculus are equivalent, given that they are all sound and complete. The Completeness Theorem for elementary classical logic says (in the so-called weak version) that all (classical) logical laws are theorems of predicate calculus (in some variant or other: axiomatic systems, natural deduction, etc.); and (in the so-called strong version) that all logical consequences of any set of premises expressible in first-order logical language are derivable from those premises in predicate calculus. The proof of the completeness of classical first-order logic is a momentous result of contemporary logic, due (once again) to Gödel, who established it in his PhD thesis, published in 1930. And – for reasons I will not enter into here – the claim that elementary classical logic is complete is *equivalent* to the claim that any consistent set of formulas has a model.

means that there exists a mathematical structure, namely $\mathbb{M}$, which satisfies the axioms of **TNT***. What is this structure like?

To begin with, such a model also satisfies the **TNT** axioms: these are, in fact, also axioms of **TNT***, which includes them all, being an extension of **TNT**. But second, $\mathbb{M}$ also makes $\neg\gamma$ true, since $\neg\gamma$ is an axiom of **TNT***. So $\mathbb{M}$ is a structure which *falsifies* the **TNT** Gödel sentence γ. Third, $\mathbb{M}$ has a *denumerable* domain: the things inhabiting the model are "as many as" the natural numbers in the standard model $\mathbb{N}$.[24] Both $\mathbb{N}$ and $\mathbb{M}$ are models of **TNT**, making all its theorems true, but they are structurally different: as the logical jargon goes, they are not *isomorphic*.[25] This means that **TNT**, as logicians also say, is not *categorical*. This slang expresses the fact that the axioms and theorems of **TNT** are unable to uniquely represent - to "capture," so to speak - the standard model constituting, as we know, the intuitive interpretation of **TNT**. Those axioms and theorems also hold in structures different from the familiar natural numbers world the teacher told us about at school. These different structures, such as $\mathbb{M}$, are called *non-standard models of arithmetic*.

To get to grips with the importance of what is going on here, recall that omega-consistency is a stronger condition than consistency: a theory can be omega-inconsistent, but consistent. Now an omega-inconsistent system "claims" (proves) that there's a number for which a given property $\alpha[x]$ holds; but it "denies" that 0 is the number, that 1 is the number, ..., and so on. **TNT*** does precisely this: the theory

---

[24] This follows from another metalogical result, called the *Löwenheim–Skolem Theorem*. The Theorem claims that, if a given first-order theory has a model, then it has a model whose domain is at most denumerable (specifically, this is the "downward" half of the Theorem, also called Downward Löwenheim–Skolem). This fact produces the so-called *Skolem Paradox* - although some doubt the appropriateness of calling it a real "paradox": set theory can have a first-order formalization, so the Theorem entails that it has a model with a denumerably infinite domain. But, as we know from Chapter 1, set theory proves the existence of non-denumerably infinite sets.

[25] The notion of isomorphism is a technical concept of model theory, which I won't spell out in detail. One can think of an isomorphism between two structures $\mathbb{A}$ and $\mathbb{B}$ as a bijection mapping each element of $\mathbb{A}$ one-to-one to an element of $\mathbb{B}$, and such that the operations and relations holding for elements of $\mathbb{A}$ are "preserved" into corresponding operations and relations holding for the corresponding elements of $\mathbb{B}$. What interests us is that the notion of isomorphism formally captures the intuitive idea of "structural identity" between models. In particular, when two models are isomorphic the respective domains have the same cardinality.

"claims" (proves) $\neg\gamma$, since it is an axiom; equivalently, it "affirms" $\exists x \mathbf{Prf}_{TNT}(x, \lceil \gamma \rceil)$. On the other hand, when in the business of proving (G1b), that is, the second half of the First Incompleteness Theorem, I have maintained that, for any number $n, \vdash_{TNT} \neg\mathbf{Prf}_{TNT}(\mathbf{n}, \lceil \gamma \rceil)$: that is to say, $\neg\mathbf{Prf}_{TNT}(\mathbf{0}, \lceil \gamma \rceil), \neg\mathbf{Prf}_{TNT}(\mathbf{1}, \lceil \gamma \rceil), \neg\mathbf{Prf}_{TNT}(\mathbf{2}, \lceil \gamma \rceil), \dots$, etc., are all provable in **TNT**. So they are also theorems of **TNT***, since the latter is an extension of the former inheriting all its theorems.

All things considered, **TNT*** is precisely an omega-inconsistent system. Reading the formulas in English, what is claimed here is that there exists a proof of γ in **TNT**, that is, $\exists x \mathbf{Prf}_{TNT}(x, \lceil \gamma \rceil)$. However (in agreement with **TNT** on this point), for any *natural* number n, it is denied that n is the Gödel number of a proof of γ. Despite making such "claims," **TNT*** says nothing inconsistent. What's going on?

A non-standard model, such as $\mathbb{M}$, includes *numbers different from the standard natural numbers*. In such a model $\neg\gamma$, that is, $\exists x \mathbf{Prf}_{TNT}(x, \lceil \gamma \rceil)$, turns out to be true, and its reading via arithmetization is precisely: "There exists a number x, such that x is the code of a proof of γ." On the other hand, no natural number can be this mysterious x. The **TNT** numerals **0, 1, 2**, … (that is, **0, 0', 0''**, …) don't "mention," so to speak, this number. However, this unnamed guy is a citizen of the non-standard model $\mathbb{M}$, that is, it belongs to the domain of quantification.

The philosophical discrepancy revealed by this situation may be summarized as follows. Recall what has been said when we had our first meeting with Typographical Number Theory: **TNT** is not an *abstract* formal system. It has been conceived to capture with its formal apparatus the world of natural numbers, with the familiar operations on them we have learned at elementary school. Its axioms have been phrased so as to formalize in the best possible way (for a first-order theory) the Peano axioms. When one spells out the formal language and axiomatic structure of **TNT**, one has (or believes oneself to have), so to speak, a certain antecedent intuition of what numbers should be, that is, an intuitive model. One has (or believes oneself to have) some grasp of the structure $\mathbb{N}$,[26] which one tries to capture by

---

[26] Actually, this is also controversial for some philosophers of mathematics. As we shall see in the second part of the book, anti-Platonists (and especially formalists) in the philosophy of mathematics usually claim that we have no such intuition at all – that the intuition of the mythical standard model is, in fact, a myth: there is no "intellectual vision" of the realm of numbers.

means of the numerals, the syntactic characterizations, etc. However, Gödel's First Theorem entails that this is an illusion. **TNT** holds in structures that turn out to be rather different from the standard $\mathbb{N}$ – non-standard models with strange non-standard numbers – and we cannot force the universal quantifier in the theory to range only across ordinary natural numbers. The formalism, so to speak, cannot catch in its grid merely the ordinary domain of our beloved scholastic natural numbers. Once the net has been lifted on board, we discover we have caught some unpredicted fish-numbers.

11   Supernatural numbers

At this point of the story, you are probably curious to know the nature of these unexpected fish-numbers, which are different from any natural numbers, and whose existence is announced by ¬γ. In *Gödel, Escher, Bach*, Hofstadter has poetically proposed to call them *supernatural* numbers[27] (not to be confused with Conway's surreal numbers). The best way to picture them is as infinitely large numbers, that is, numbers larger than any natural.

To get a better picture, we may take a small detour from the Incompleteness Theorem. The existence of non-standard models for **TNT** follows, in fact, also independently of G1, for it is entailed by a key feature of elementary formal systems, expressed by the *Compactness Theorem* for first-order logic. The Compactness Theorem can be expressed, for instance, by saying that a first-order formula is a consequence of a given set of first-order premises if and only if it already follows from a *finite* subset of those premises; and this is equivalent to the claim that a set of formulas has a model if and only if all of its finite subsets have. We are interested not in the Theorem itself, though (which is why I am dealing with it quickly), but only in the fact that, by means of it, one can prove the existence of non-standard models as follows.[28]

---

[27]   See Hofstadter (1979), Ch. XIV.
[28]   Compare Franzén (2005), pp. 135–6.

First, add to the **TNT** language an individual constant, say s (for "supernatural"). Next, consider the following theory – call it **TNTs** – obtained by adding to **TNT** an infinite list of axioms such as the following:

$(\text{TNT}_s 1)$	$0 < s$
$(\text{TNT}_s 2)$	$1 < s$
$(\text{TNT}_s 3)$	$2 < s$
$(\text{TNT}_s 4)$	$3 < s$
$(\text{TNT}_s 5)$	$4 < s$
$\ldots$	$\ldots$

and so on.[29] It is easy to show that any finitely axiomatized subtheory of **TNTs** has a model in $\mathbb{N}$. By the Compactness Theorem, then, **TNTs** also has a model, which is precisely a non-standard one. Besides including the numbers denoted by the numerals $0, 1, 2, \ldots$ (that is, as usual, $0, 0', 0'', \ldots$), which correspond to the standard naturals, such a model includes a number, denoted by s, that has to be larger than *all* the predecessors. The infinite list of axioms gives us a (roughly) intuitive idea of the supernatural number (denoted by) s being *infinitely large* with respect to the naturals.

What are we to do with these non-standard models and their supernatural denizens? An articulated answer to this question would straightforwardly carry us into the controversial domain of philosophy, or at least of the philosophy of mathematics. Some people believe such models to be intrinsically "pathological." For instance, some take them as confirming criticisms advanced by the intuitionists against Hilbert's formalist idea that the syntactic consistency of a formal system is sufficient to grant it a model. Consistency – so the objection goes – is not enough to present an authentic mathematical *reality* which would be described by the relevant system. The only thing non-standard models actually illustrate is that the consistency of a theory is not sufficient for it to avoid "making nonsensical claims." Once supplied with an interpretation, the theory deals with models having nothing to do with the good old teacher's arithmetic.

---

[29] Recall that "<" can be defined in **TNT** by means of + and the logical symbols.

In the eighth chapter of this book I shall talk of the strange bifurcations of Typographical Number Theory supplied by the non-standard models, and of their alleged "postmodern" consequences. We will take into account the claim that, when our famous good old teacher propounded truth in the standard model ℕ as the Absolute Arithmetical Truth, she was not being that innocent after all. However, there also exists an uncontroversial positive mathematical answer to the question concerning the value of non-standard models in general. It is easy to have non-standard models of the real numbers, including "infinitely small" and "infinitely large" numbers. Since the sixties, non-standard mathematical structures have been at the core of *non-standard analysis*, which has allowed us to make consistent and develop the infinitesimal calculus due to Newton and Leibniz. When mathematical logicians (who usually get the smallest offices in mathematics departments) want to boast that their discipline makes concrete contributions to mathematics, they usually mention the case of non-standard analysis.

12   The culpability of the induction scheme

It is often said that the primary responsibility for the incompleteness and non-categoricity of **TNT** rests with the axiom scheme of induction.[30] Such a scheme, in fact, is just a simulacrum of the strength of real mathematical induction. Recall what the original, informal Peano axiom claimed:

(P5)   Any property of zero that is also a property of the successor of any number having it is a property of all numbers.

This means that any set of natural numbers containing 0 and closed under the successor operation contains all natural numbers. But, as we now know, first-order logical languages such as the one **TNT** is phrased in cannot express such things as "all properties" and "some properties":

[30]   See Smullyan (1992), p. 112.

in the specific case, they cannot quantify on sets of natural numbers. Now, by Cantor's Theorem (the power set of a given set is larger than the set itself), the set of all subsets of the naturals is *more* than denumerable, so it is larger than the set N of natural numbers. There are, therefore, more than denumerably many properties of the naturals. But there are at most denumerably many **TNT** formulas to replace the metavariable α in the **TNT** induction scheme; so the properties of natural numbers expressible in the **TNT** language are at most as many as the naturals.

Now the thought might occur that, although the incompleteness of **TNT** *as* a first-order formal system is essential and irredeemable, things may improve if we phrase the Peano axioms (specifically the fifth) in a higher-order language allowing quantification on properties and, in the specific case, on all the properties of natural numbers. Such a language would include, besides the individual variables x, y, ..., also predicate variables – say X, Y, ..., etc. – which can occupy in the language the places of predicate constants (just as individual variables can take the place of individual constants), and be bound by quantifiers. The "real" translation of (P5) in the enriched language, then, would go as follows:

$$\forall X(X(0) \land \forall x(X(x) \to X(x')) \to \forall x X(x)).$$

Now, second-order Peano arithmetic is *categorical*, that is, it does not have odd non-standard models.[31] When we move to a second-order language it seems we can, in a sense, pick the model N uniquely ("up to isomorphism," as is usually said). However, the problem of incompleteness is *not* solved at all. Even though mathematical induction has a full second-order recapture, now the problems stem from second-order *logic*, which is not axiomatizable. Elementary logic, that is, the classical first-order predicate calculus with identity, allows us to deduce all the logical consequences of the (first-order) premises we assume. But this is not the case with second-order logic: no formalization of second-order logic is complete in this sense. Furthermore, second-order logic is often accused to be (quoting Quine) "set theory in sheep's clothing" in its turn; and mainstream set theories also fall within the range of the formal systems to which the Incompleteness Theorem

---

[31]   For a proof of this, see e.g. Moriconi (2001), pp. 176 ff.

applies. Also from a conceptual viewpoint, therefore, resorting to second-order logic in the hope of resolving the problem of incompleteness is like moving in a small circle.

13   Bits of truth (not too much of it, though)

To conclude the first part of our book, something needs to be said on the issue of the *truth* of γ (and of ¬γ). Something – but not too much, for when truth enters the scene philosophical argy-bargy begins, and this is properly postponed to the second part.

When proving G1 and G2, I have followed as closely as possible the spirit of the original Gödelian proof, and I have especially wanted to stick to the "syntactic" level, avoiding the "semantic" one in a broad sense. Of course, I have often referred to the "reading" of the **TNT** formulas: I've been talking of a formula being "readable" (by means of the arithmetization of syntax) as saying that …, and so on. But these were only shorthand ways of making the following claim: such-and-such formula formally represents (in the precise sense of *formal representation* we defined some chapters ago) within **TNT** an arithmetical statement; and this statement in its turn is the counterpart, obtained via the Gödel numbering (that is, via a *calculus*), of a metamathematical statement on **TNT**. The abbreviations I have adopted for the relevant arithmetic notions, such as "Prf_{TNT}" (for the intuitive), "$\mathbf{Prf}_{TNT}$" (for the formal), and so on, have been chosen in order to help us to bear in mind the metamathematical correlates: to help us bear in mind that Prf_{TNT} is the arithmetic relation associated by means of the arithmetization procedure with the metamathematical-syntactic proof relation of the system, that is, that Prf_{TNT} holds between those pairs of numbers m and n (the proof pairs), such that m is the code of a proof of **TNT**, and n is the code of the proved formula; that $\mathbf{Prf}_{TNT}$ is the formal correlate of Prf_{TNT}, that is, the expression "$\mathbf{Prf}_{TNT}$" is the two-place predicate formally representing that relation in **TNT**; and so on.

I have also insisted on the fact that the semantic assumption of the soundness or correctness of **TNT** (that is, its proving only *true* things) has not been used in the proof of G1 and G2. And I have maintained that Gödel carried out the details of his amazing proof by adopting the notion of (omega-) consistency precisely to avoid resorting to the notion of mathematical truth. G1 and G2, as formulated above, are syntactic results in this sense. Specifically, G1 attests only to the syntactic incompleteness of **TNT**, which means (putting things down as formalistically as possible): assuming the (omega-) consistency of **TNT**, there exists a string of symbols, which we have dubbed γ, not producible as a theorem of **TNT**, and such that the same string with a "¬" put in front of it is also not producible as a theorem. Or, given the axioms of **TNT**, which are finite strings of symbols, and the rules of inference, which are instructions for the manipulation of symbols, if **TNT** is (omega-) consistent, then neither the string called γ nor the string called ¬γ will ever show up in the last line of a proof of **TNT**, that is, in the last line of a sequence of strings obtained starting with the axioms, and by means of manipulations allowed by the rules.

This said, when examining the "immediate consequences" of G1 and G2, I have been forced to talk in a systematic way of *models* – mathematical structures making the theorems of **TNT** *true* – and in doing so I was just sticking to the standard practice in mathematical logic. I have been talking of the fact that we might have some intuitive grasp of the realm of natural numbers, (represented by) the celebrated ℕ, which **TNT** was supposed to formally capture. It has even been claimed that there exist strange mathematical universes (the non-standard models) which falsify γ, and witness against the hope of a univocal capture of ℕ by means of **TNT**. It has been claimed, thus, that a formula is *true in* some model, whereas its negation is *true in* some other. Relativizing truth in this way, of course, is mandatory: for otherwise, if γ and ¬γ were declared to be both true *simpliciter*, or true in the same model, we would be violating the Law of Non-Contradiction. So now one might be wondering (especially if one comes from a philosophical background) what can ultimately be said about the truth of the Gödel sentence and of its negation. If it does make sense to speak of their truth and falsity at all, which is the true one – the *truly* true one?

Notice that the syntactic incompleteness of **TNT** produces a semantic incompleteness if we accept that the Principle of Bivalence, stating that any sentence is either true or false (i.e., such that its negation is true), holds for the **TNT** sentences once they have been interpreted. As you may recall, what I have called semantic incompleteness some chapters ago is the fact that there exists at least one true formula which is unprovable in **TNT**. For if Bivalence holds, then *one* of γ and ¬γ has to be a true statement, but **TNT** proves neither.

However, does Bivalence actually hold in general for mathematical statements? Some philosophers of mathematics answer in the negative. And then again, which one between γ and ¬γ would be the true one in any case – the *truly* true one?

When such questions begin to show up, the "immediacy" of the consequences of Gödel's Theorem begins to fade. By resorting to such concepts as truth, Bivalence (for mathematical statements), interpretation (of the **TNT** formulas), mathematical intuition, we begin to move into the uncertain ground of the philosophy of mathematics. Soon we find ourselves in the territory of metaphysics: we begin to posit questions on what truth is, on the ontological status of numbers, and so on.

And, of course, now that the going gets tough, the philosophers get going.

Part II
The World after Gödel

Replying to your inquiries I would like to say first that I don't consider my work "a facet of the intellectual atmosphere of the early 20ᵗʰ century" but rather the opposite. (Kurt Gödel, unsent letter to the sociologist Burke D. Grandjean)

8

Bourgeois Mathematicians! The Postmodern Interpretations

Armchair philosophy has a poor reputation among many scientists. One of the reasons is that some philosophers have the habit of commenting on the latest and most esoteric achievements in physics, mathematics, biology, etc. – showing with their remarks that they understand little of the specialized results they talk about. Scientists, physicists, biologists, or mathematicians are often perplexed by the extreme arbitrariness of the philosophical interpretations of their own efforts; it is as if they were doing the hard work, just to see the philosophers come on stage at the end of the day and teach them the True Meaning of what they (the scientists) had been doing all along. Things are especially bad with so-called "continental" philosophy, for during the last century practitioners of this subfield elaborated theories to the effect that producing free interpretations of almost anything is *good* philosophical practice. And with Gödel's Theorem they have been extremely productive. Perhaps just to be loyal to my party, I have a strong tendency to support philosophers – with a few exceptions. As we shall see in this second part of the book, problematic theoretical arguments against artificial intelligence have been proposed by resorting to the Incompleteness Theorem; and illustrious philosophers have advanced interpretations of the Theorem whose significance is quite controversial among the philosophers of mathematics. But in these cases, even granting that the arguments at issue are fallacious, they are so in an interesting way. The proposed readings are interesting – or so they seem to me – in the following sense: their very refutation reveals fascinating aspects of the Theorem, and helps us understand better its spinoffs. I do not believe that the same thing could be claimed of the interpretations provided by much postmodern thought. Let us see why.

1 What is postmodernism?

What is postmodernism, to begin with? Here is a characterization by Eagleton:

> Postmodernism is a style of thought which is suspicious of classical notions of truth, reason, identity, and objectivity, of the idea of universal progress or emancipation, of single frameworks, grand narratives or ultimate grounds of explanation … [It] sees the world as contingent, ungrounded, diverse, unstable, indeterminate, a set of disunified cultures or interpretations which breed a degree of skepticism about the objectivity of truth, history, and norms, the givenness of natures and the coherence of identities.[1]

The label "postmodern" is attached to a variety of philosophical schools, often quite different from one another: from French deconstructionism, to what is called "weak thought" in Italy, to some branches of philosophical hermeneutics along the line of Nietzsche, Heidegger, and Gadamer. A common trait of these schools is that, whereas on the one hand the history of philosophy gets a lot of attention (much more than in the so-called "analytic philosophy"), on the other hand postmodern philosophers criticize the philosophical tradition because of its attempts to provide a definitive foundation for thought, its quest for firm, unassailable truths, for the "absolute knowledge," etc. The postmodern reaction is often summarized in such mottos as Nietzsche's "There are no facts, only interpretations." It develops in a variety of criticisms to such notions as objectivity and objective (unhistorical, not to say incontrovertible) truth. In this framework, there is no fact of the matter concerning how things stand in themselves, in more or less any field of inquiry: truth *is* not, truth is *made*, for nothing subsists by itself, independently of human beings' interpreting and meaning-conferring activity. There is no objective, given reality; so *a fortiori* there are no theories capable of getting to grips with such reality; those who attempt such theories are labeled "metaphysical" – where the term carries the negative connotation of a pretentious, reductive, even *violent* style of thought.

---

[1] Eagleton (1996), pp. vii–viii.

Sometimes postmodernism goes hand in hand with philosophical *relativism* – a position with some 2,000 years of history, which can be characterized very plainly as the claim that truth *simpliciter* makes no sense. What does make sense is a relativized notion of truth: one can claim that something is true in a particular historical setting, or given some (always revisable) presuppositions, social conventions, and so on.

We are not interested in the niceties of philosophical postmodernisms, though. What does interest us is the question: what does Gödel's Theorem have to do with all this?

Postmodern interpretations of the Incompleteness Theorem are often linked to skeptical conclusions – that is, to claims of epistemic flavour. The main thought is that Gödel's result corroborates the idea that no definitive, incontrovertible truth is available, by entailing that we cannot reach definitive *mathematical* truths. This sounds like a momentous philosophical outcome, since mathematics, taken as the field in which something definitive had indeed been established, was the citadel of resistance for rationalists. However, I plan to talk at length about these issues in Chapter 10, which is devoted to the skeptical interpretations of the Incompleteness Theorem, and where epistemic conclusions of a skeptical kind will be argued for more persuasively.

For now, I will deal with two other typically postmodern readings of the Theorem: (1) that extending its implications to extra-mathematical and extra-logical contexts, having incompleteness invade metaphysics, physics, the social sciences, and even religion; and (2) that using the Theorem to testify the inexistence of any objective mathematical reality (and, therefore, truth) or, at the very least, the necessity of relativizing the concept of mathematical truth.

2   From Gödel to Lenin

Alan Sokal and Jean Bricmont have dealt with the former postmodern reading quite convincingly in their bestseller *Intellectual Impostures*. Here they have attacked the logically and scientifically preposterous claims of such philosophers as Gilles Deleuze, Félix Guattari, Julia Kristeva, and Luce Irigaray. Régis Debray is probably the greatest producer of ridiculous claims on Gödel's Theorem:

Ever since Gödel showed that there does not exist a proof of the consistency of Peano's arithmetic that is formalizable within this theory (1931), political scientists had the means for understanding why it was necessary to mummify Lenin and display him to the "accidental" comrades in a mausoleum, at the Center of the National Community.[2]

In Debray's *Critique of Political Reason* it turns out that collective madness "has its ultimate foundation in a logical axiom, itself unfounded: *incompleteness*":[3]

> The statement of the "secret" of collective misfortunes, that is to say, of the *a priori* condition of all political history, past, present and to come, is to be found in a few words, simple and childish … This secret has the form of a logical law, a generalization of Gödel's theorem: there is no organized system without closure, and *no system can be closed with the help only of the elements belonging to the system.*[4]

The philosopher Michel Serres has identified a "principle of Gödel–Debray," explaining how Régis Debray

> applies to social groups or finds in them the theorem of incompleteness, which holds for formal systems, and shows that societies only organize themselves on the express condition that they are founded on something other than themselves, external to their definition or boundary. They cannot be self-sufficient. He calls this foundation religious. By way of Gödel he completes Bergson.[5]

The first reason why such "generalization of Gödel's theorem" is outlandish and there is no "Gödel–Debray principle" at all is quite simple: there exists no connection whatsoever between Gödel's Theorem and social organizations. The same holds for similar attempts to extend incompleteness to juridical, political, or religious contexts. Here is a laughable quotation reported by Franzén:

> Religious people claim that all answers are found in the Bible or in whatever text they use. That means the Bible is a complete system, so Gödel

---

[2]   Debray (1980), p. 70.
[3]   Debray (1981), p. 10.
[4]   Ibid., p. 265.
[5]   Serres (1989), p. 358.

seems to indicate it cannot be true. And the same may be said of any religion which claims, as they all do, a final set of answers.[6]

The problem with statements of this kind – religious, sociological, etc. – is that, as we have abundantly learned, Gödel's (First) Theorem claims that certain formal systems capable of representing a portion of elementary arithmetic include undecidable formulas, and therefore they are incomplete. Taken as a syntactic result, as a pure formalist would, this means only that if the systems at issue are consistent (or omega-consistent), then some strings of symbols and their respective negations are not theorems of the systems, …, and all that. Once we have supplied such strings with an interpretation, we know that they are (formal representations of) arithmetical claims on numbers. Within the so-called human sciences, as well as in political or religious debates, one can hear talk of "incompleteness," "consistency," "indecision," etc. In itself, though, this is not sufficient to substantiate any interpretation according to which those arithmetic formulas may talk in some illuminating way of political systems, incomplete societies, psychological indecisions, or consistency in politics.

3   Is "Biblical proof" decidable?

There's another obvious reason why Gödel's Theorem cannot be applied to political theories, sociological analyses, legislative corpuses, Marx's *Capital*, the Koran, or the Bible: in all these cases, we are not dealing with *formal systems* at all. First, these "theories" in a broad sense are phrased not in a formal language, but in some natural language or other. Even if we translated them into some logical formalism, most devotees of such texts as *Capital*, the Koran, or the Bible would protest that in doing so we had altered their nature. But even if the translation were generally accepted, still an organization in terms of axioms and theorems would be missing. And even if the communists, the Muslims, and the Christians found some general agreement on which are the axioms of *Capital*, the Koran, and the Bible respectively,

---

[6]   See Franzén (2005), p. 77.

and on which logic is actually used by Marx[7] or by God to delineate the consequences of those axioms (is God a classicist or an intuitionist?), there would remain the problem of showing that the set of such axioms, as well as the notions of *Biblical proof*, *Leninist proof*, etc., are decidable.

As we also know, a theory like the True Arithmetic is semantically complete, in the sense that no arithmetical truth escapes it, simply by definition. But it escapes from the Incompleteness Theorems precisely because it doesn't fulfill the Fundamental Property of formal systems (and therefore, putting it the other way around, it does not escape from the Theorems at all, since Gödel never wanted them to apply to anything but formal systems, that is, formal theories satisfying the Fundamental Property). The same should be said, *a fortiori*, of *Capital*, the Koran, and the Bible. In a sense, it had better be so. If the notion of biblical proof were recursive, and the biblical theorems were recursively enumerable, we could instruct a computer to print them systematically, thus leaving the exegetes with little work to do.

Next, are the Bible, the Koran, or *Capital* consistent? One could claim that, when such terms as "consistency," or "soundness," or "completeness," are applied to texts which are neither formal systems nor formalized theories, such as the books at issue, they should be interpreted by analogy with respect to the technical sense the terms have in logical and mathematical contexts. Specifically, one might say that they have a fairly precise sense when read semantically. After all, semantic completeness has to do with telling all the truth, and soundness has to do with telling only the truth – two things we expect also from witnesses in a trial. Do the Bible, or *Capital*, etc., tell us all the truth on the things they talk about (their "domain," or "universe of discourse"), then? One may also wonder whether the Bible or *Capital* are simply consistent, in that these books don't claim somewhere things they have denied somewhere else. Now it is likely that the Bible, or *Capital*, are inconsistent in this sense, and it is virtually certain that they are incomplete: just as Tolkien's *The Lord of the Rings* doesn't tell us whether Pippin Took had more than 96,000 hairs on his head or on which day fell his fifteenth birthday, so the Bible doesn't tell us whether Solomon had

[7] Of course, Hegel and Marx adopted *dialectical* logic, but dialectical logic hasn't much to do with formal systems in the modern sense: on this issue, see Berto (2005), Chs 2 and 3.

more than 96,000 hairs on his head or on which day fell his fifteenth birthday. As Franzén has claimed, "we don't need Gödel to tell us that these 'systems' are in this sense incomplete. Trivially, any doctrine, theory, or canon is incomplete in this analogical sense."[8]

4 Speaking of the totality

The point just made holds also for more interesting attempts to export incompleteness in extra-mathematical contexts – particularly, in physics and metaphysics.

Philosophy undergraduates are told that philosophy begins when one has some Insight of the Totality. Western philosophy starts when the ancient Greeks began to posit questions on the *arché* of all things, thereby assuming that it makes sense to talk about the totality of things – about the world as a whole. Specifically, philosophy looks like a Theory of the Whole in so far as it is taken in the sense of traditional metaphysics. This was defined by Aristotle as the science which studies being *qua* being – and the properties being has *qua* being – and therefore as the discipline that deals with *everything*.[9] This does not mean, of course, that it deals with each single thing: it means that metaphysics isn't the study of a specific area of reality (the physical world, or the psychological one, or the realm of numbers), and of the properties some things have *as* things of a specific kind. Metaphysics deals with the most general features of reality – with the properties things have in so far as they are things. But Aristotle's definition may apply as well when translated in terms of contemporary analytic metaphysics. When Quine claims in "On What There Is" that the basic ontological question is, "What is there?," he maintains that ontology has to do with the Whole – with what exists, with the furniture of the world. To be there, or to exist, in this kind of jargon means nothing more and nothing less than to belong to the Whole. Now, if the basic intuition that leads to philosophy is the Insight of the Totality, one may wonder whether metaphysics, taken as the Theory of the Whole, is rendered impossible by Gödel's Incompleteness Theorem.

---

[8] Franzén (2005), p. 79.
[9] As is well known, Aristotle didn't use the word "metaphysics"; furthermore, what is characterized as metaphysics here has also been called "general ontology."

Franzén remarks that a similar question has been posited for what is sometimes called the Theory of Everything. This is something like an ideal theory of theoretical physics, from which all of the physical truths could be inferred. Physicists of philosophical inclination have proposed various drafts of a Theory of Everything, but the experimental basis of such draft theories is quite dubious. It is particularly difficult to have them match with the descriptions of the universe provided by the two mainstream theories around, that is, quantum mechanics and general relativity, which are hardly compatible.

This said, a physical Theory of Everything would be itself a metaphysical Theory of the Whole, from some reductionist philosophical viewpoint. If someone believes that reality is just physical reality and, therefore, that general ontology should become one with (or maybe should defer to) general physics, then a hypothetical Theory of Everything may end up identified with the metaphysics of the Whole: something whose theorems settle all issues in physics (and therefore, in the last instance, all issues on what there is), and exhaust in principle the totality of physical truths (and therefore, in the last instance, the totality of true claims on what there is).

Now the authoritative physicist Freeman Dyson, engaging in philosophical speculation, has written:

> Another reason why I believe science to be inexhaustible is Gödel's theorem … his theorem implies that pure mathematics is inexhaustible. No matter how many problems we solve, there will always be other problems that cannot be solved within the existing rules. Now I claim that because of Gödel's theorem, physics is inexhaustible too. The laws of physics are a finite set of rules, and include the rules for doing mathematics, so that Gödel's theorem applies to them. The theorem implies that even within the domain of the basic equations of physics, our knowledge will always be incomplete.[10]

Is the idea of an all-inclusive and exhaustive theory of reality refuted by the Incompleteness Theorem, then? Suppose we can arrange a hypothetical metaphysical Theory of the Whole, or a hypothetical physical Theory of Everything (which according to some may turn out to be the same, as we have seen), in an authentic formal system: something whose theorems are recursively enumerable. Such a theory may well include

---

[10]   Quoted in Franzén (2005), p. 87.

arithmetical axioms. In this case the theory, as Franzén has pointed out,[11] would be syntactically incomplete in its arithmetical fragment: there would be arithmetical statements expressible in the language of the theory, which the theory could neither prove nor refute. If Bivalence holds for the realm of arithmetic, then the theory would be also semantically incomplete: some arithmetical truth would be left out of the set of theorems, so the theory would not be exhaustive in this sense. Then again, this does not entail that any *non*-mathematical, and specifically physical, incompleteness has been established. Even though the theory cannot decide any arithmetical statement with its deductive apparatus, this does not lead to incompleteness in the description of the physical world. This fact too, Franzén concludes, "is not something the incompleteness theorem tells us anything about."[12]

However, these negative conclusions point at something positive. It has been claimed that physical science or metaphysics should be incomplete because of Gödel's Theorem. The problem with these statements, as we have seen, is that the Theorem is a result in mathematical logic, and difficult to sell abroad in non-mathematical contexts. One may conjecture that the Theorem shows, nevertheless, that *mathematics* is inexhaustible. I think it does show this, in a sense. But which sense? As often happens in philosophy, to get to grips with a key theoretical point one has to fine-tune the meanings of some words; in this case, what stands in need of clarification is the term "inexhaustible." I shall talk about this at length in Chapter 11, which is devoted to the alleged consequences of Gödel's Theorem in artificial intelligence and the philosophy of mind.

5   Bourgeois teachers!

The second postmodern interpretation flagged above maintains that the Incompleteness Theorem bears witness, if not to the inexistence of any mathematical reality, at least to the necessity of relativizing the concept of mathematical truth. This second reading is more complex to deal with, and the problems it raises are subtler. Let us start with a

[11]   See ibid., pp. 87–8.
[12]   Ibid., p. 88.

retrospective look at the development of the first part of this book with respect to the notion of mathematical truth. I started by just talking of arithmetical truth and falsity. I then introduced the notion of formal system: a specific system, namely our beloved Typographical Number Theory, and its intuitive interpretation – the realm of natural numbers our trustworthy teacher told us about when we were children. After providing bits of model theory, I claimed that the intuitive interpretation is in fact (represented by) a model, the legendary ℕ. But the First Incompleteness Theorem entails that **TNT** has non-standard models too.[13] So I found myself saying that some strange **TNT** formulas can be *true in* some model, and *false in* some other. I finished the first part of the book with an open question on which sentences are true *simpliciter* – truly true.

Now the second postmodern interpretation of Gödel's (First) Theorem consists in claiming that this question makes no sense. What the (First) Theorem shows is that there simply is no unique, objective arithmetical reality we can truly grasp (or fail to grasp, by the way). Here is a sample quotation from William Barrett's *The Irrational Man: A Study in Existential Philosophy*, where Gödel joins Nietzsche and Heidegger in the club of the Destructors of Objectivity:

> In the Western tradition, from the Pythagoreans and Plato onward, mathematics as the very model of intelligibility has been the central citadel of rationalism. Now it turns out that even in his most precise science – in the province where his reason had seemed omnipotent – man cannot escape his essential finitude: every system of mathematics that he constructs is doomed to incompleteness. Gödel has shown that mathematics has insoluble problems, and hence can never be formalized in any complete system ... Mathematicians now know they can never reach rock bottom; in fact, there is no rock bottom, since mathematics has no self-subsistent reality independent of the human activity that mathematicians carry on.[14]

Some forms of postmodernism maintain that the so-called objective reality is but our dominant convention: if we had different mainstream

---

[13]   You might recall, though, that these also follow independently of the First Theorem, on the basis of the compactness of elementary logic: we would know about the existence of non-standard models even if Gödel had never published his result.

[14]   Quoted in Goldstein (2005), p. 39.

conventions, our so-called objective reality would be different. In a sense, it does not even make sense to speak of objective reality anymore: selling our dominant convention as a set of objective facts is just ideological propaganda. Analogously, the incompleteness of such systems as **TNT** is taken as showing that the so-called objective reality of numbers is but our favorite arithmetic. As testified by the non-standard models, we could adopt different arithmetics in which γ is refuted. We could have mathematical realms whose populace includes non-standard numbers quite different from those our educators told us about at school. The good old teacher, after all, was a bourgeois ideologist!

Mathematicians may turn out to be bourgeois too. Of course, working mathematicians normally make a naïve and, so to speak, "naïvely deflationist" use of the truth predicate (more or less as most of us were doing before attending some philosophy course which disturbed our natural intuitions). When the next man says that it is true that snow is white, he means that snow is white. And when a mathematician sets out to prove that Fermat's conjecture is true, she aims at showing that there are no solutions in positive integers for $x^n + y^n = z^n$ when $n > 2$. According to the postmodern framework outlined above, however, when mathematicians cheerfully talk of the proof of some arithmetical theorem or other, meaning that it has been established as true, they might be involuntarily engaged in ideological talk by omitting the qualification: true *in* the standard model $\mathbb{N}$, that is, in the dominant mathematical ideology.

6   (Un)interesting bifurcations

This may still sound like postmodern rubbish. However, the situation can be made more insidious by means of a few technical considerations. Consider the "multifurcations" of Typographical Number Theory that Hofstadter talks about in *Gödel, Escher, Bach*. We already know a first **TNT** bifurcation – we met it at the end of the first part of this book. Since **TNT** is syntactically incomplete, we can *extend* it in two different directions: $\mathbf{TNT}_1 = \mathbf{TNT} + \gamma$, and $\mathbf{TNT}^* = \mathbf{TNT} + \neg\gamma$. $\mathbf{TNT}_1$ is the *consistent* theory obtained by adding to **TNT** its Gödel sentence as an axiom, and **TNT*** is the consistent theory obtained by adding to **TNT** the negation of the Gödel sentence. We also know that the process of extending $\mathbf{TNT_1}$ into $\mathbf{TNT_2}$ ($= \mathbf{TNT_1} + \gamma_1$), $\mathbf{TNT_2}$ into $\mathbf{TNT_3}$ ($= \mathbf{TNT_2} + \gamma_2$), and so on, can be protracted indefinitely by adding to each new system its Gödel sentence. By proceeding in this way, we keep ourselves on the standard side of the available extensions. But we could also walk along the non-standard side, by adding the negations of the various Gödel sentences. We might even zigzag, by taking a standard direction and then turning to a non-standard one: for instance, we could pick $\mathbf{TNT^*}$ (that is, $\mathbf{TNT} + \neg\gamma$) and add to it its Gödel sentence, say γ^*; and so on. In these contexts, people sometimes put forth analogies with *non-Euclidean geometries*. Let us have a look.

In the case of the Peano axioms for arithmetic, we had an intuitive grasp of the meanings of the terms "zero," "number," "successor," which the Peano axioms systematized. Analogously, with Euclidean geometry there must have been some intuitive understanding of the meanings of such terms as "line," "point," "plane," etc. And Euclid's *Elements*, as advertised at the beginning of this book, managed to account for these notions by means of some definitions and axioms, which appeared to be so intuitively plausible that they were taken as obviously true – true *simpliciter*, that is: as belonging to a theory that was to be the One True Description of the One True Space. Euclid's lines were taken as *the* lines, Euclid's planes were taken as *the* planes (Kant believed this, and he is often shamefully blamed for having held such a credo).

However, as centuries passed, doubts started to accumulate on the fifth and last postulate. The other four consisted of simple phrases such as "Any two points can be joined by a straight line." But the fifth sounded like: "If two lines intersect a third in such a way that the sum of the inner angles on one side is less than two right angles, then the two lines inevitably must intersect each other on that side if extended far enough." The qualms probably had their roots in psychology, or even aesthetic chauvinism (but beauty is a guide to truth, as they say): this fifth principle is uglier than the others, too long and roundabout in its formulation. Initially, people wondered not whether it was true or false, but whether it was a theorem rather than an axiom (that is, actually independent, not deducible from the others). Of course, if one managed to infer it from the others, then one could get rid of this fleshy guy. Various alleged proofs were produced, but they all included some mistake or other. The strategy inaugurated by the early-eighteenth-century Italian mathematician Girolamo Saccheri consisted in bracketing the postulate, or more precisely, in assuming a specific principle incompatible with an equivalent formulation: "For any given line *l* and point *P* not on *l*, there is exactly one line through *P* that does not intersect *l*." The idea was that, if the fifth postulate is a real axiom, then its negation does not contradict the other four, being independent. But Saccheri thought the consequences of such an operation to be "repugnant to the nature of straight lines." If one believes that our intuitions on the meaning of "straight line" provide us with *the* model which refutes any negation of the fifth postulate, one should stop here. So did Kant: believing Euclidean geometry to be the One True Geometry, he made of it the science based upon the pure intuition of space: One Pure Intuition for the One True Space.

Nowadays we know that the fifth postulate is actually independent from the others. Therefore, theories assuming, in addition to the first four principles, some postulate which entails the denial of the fifth, are themselves consistent. The belief that any consistent theory must have some model leads us to non-Euclidean geometries: theories that keep the first four Euclidean axioms (often called *absolute* or *neutral* geometry), but dispense with the fifth by assuming that there is more than one parallel to the famous straight line, or none, and are called, respectively, hyperbolic and elliptic geometry.[15] Now we may dispense with the persuasion that our intuitive model is the only one, and let the meanings of "line," "surface," etc., be *determined by the theory itself*. Theories can, so to speak, act backwards on our intuitions. We discover that the first four axioms can receive alternative meaningful interpretation, that is, they have (initially) counter-intuitive but actual deviant models. And non-Euclidean geometries have many interesting and useful applications.

A difference with our Typographical Number Theory is that, in the case of **TNT**, there is an *infinity* of available bifurcations, all consistent, all with their model(s). These are infinitely many mathematical structures, variously incompatible with each other, but all satisfying the original **TNT**. We have begun with the hope of fully formalizing intuitive arithmetic by means of **TNT**; we have discovered that **TNT** is not categorical, and catches unexpected fish in its formal net. And now, here come the infinitely numerous mathematical universes beyond the (ideological) standard model ℕ, whose existence is announced by the

---

[15] For the latter one also has to drop the Euclidean principle that a line may be extended indefinitely.

model-theoretic upshots of Gödel's First Theorem. Instead of sticking to our intuitions on the presupposed (or stereotypical) model, how about letting mathematical structures be unfolded by the infinitely many theories producible by extending **TNT**? Could there be anything of a more postmodern flavour? Let us listen to J. Kadvany on this point:

> The simplest observation of how Gödel's Theorems create a postmodern condition begins with the First Incompleteness Theorem … Since an undecidable proposition and its negation are separately consistent with the base system, one can extend the old system to two mutually incompatible new ones by adding on the undecidable sentence or its negation as a new axiom … The new systems so constructed also have new undecidable sentences, different from the originals, and the process of constructing new undecidable sentences and then new systems incorporating them or their negations goes on *ad infinitum*, like a branching tree which never ends.[16]

Franzén's reply consists in stressing that one can extend incomplete arithmetical systems in various ways which have no mathematical interest whatsoever. Take the third axiom of **TNT**:

(TNT3) $\forall x(x + 0 = x)$.

This is the formal counterpart of the apparently obvious arithmetical claim: "By adding 0 to any given number, we still have that number." This axiom is independent from the others, as good axioms in a formal system are expected to be: it cannot be inferred from them. Suppose we delete it from **TNT**. In the **TNT** so mutilated, (TNT3) becomes an undecidable sentence. Now we can give to our theory this other axiom in return, that is, the negation of the previous one:

(NonTNT3) $\exists x \neg (x + 0 = x)$.

(NonTNT3) "claims" that there exists some number x, such that adding 0 to it gives something different from x. Call **NonTNT** the theory thus obtained. **NonTNT** is consistent in its turn; therefore, as usual, it has a

---

[16]   Kadvany (1989), p. 162.

model. Among the denizens of such a model there must be some guy – call it q – such that $q + 0 \neq q$. If we followed the postmodern line of thought now, we might come up with such claims as: "To say that when one adds 0 to any number whatsoever one obtains the same number is just ideology. It is time to realize that such a statement is true only *in some sense* ....All in all, there are numbers different from the usual natural numbers, like q, such that adding zero to them gives something different from q – and we don't want to discriminate against them, do we?"

After listening to our postmodern philosopher, most of us will just kick this q away, and stick to our persuasion that adding zero to any number gives just that number. But do considerations of this kind entail that "The incompleteness theorem has not led to any situation in which mathematics (and specifically arithmetic) branches off into an infinity of incompatible directions?"[17] It seems to me that at least some of these bifurcations are not that pathological. First, I have mentioned that non-standard models are at the core of contemporary non-standard analysis, and this is a lively, well-established, and expanding branch of mathematics. It is also to be granted that the existence of non-standard models is not tied exclusively to the Incompleteness Theorem, since it follows also by way of Compactness (and True Arithmetic has non-standard models too).

Second, it is not so clear that any deviation from standard arithmetic obtained by doing without allegedly obvious truths such as (TNT3) is pathological and uninteresting. Towards the end of this book, we shall discover the existence of certain non-standard arithmetic structures (the paraconsistent arithmetics) which dispense to some extent with such truths as that each number is distinct from its successor: they include some number n such that $n = n + 1$! However, as we will see, these strongly deviant theories display decidedly interesting mathematical and metamathematical features.

---

[17] Franzén (2005), p. 54.

9

A Footnote to Plato

The co-author of *Principia mathematica* Alfred North Whitehead claimed that the whole European philosophical tradition can be fairly characterized as a series of footnotes to Plato. If this is true, then one of these notes – an important one too – was signed by Kurt Gödel. There exists a Platonist interpretation of the Incompleteness Theorem and, in fact, this is a self-interpretation: it is the reading provided by its author. This in itself is a good reason to take the interpretation seriously from the beginning. To understand it, we should at the outset learn something about the nature of Platonism in the philosophy of mathematics, and on why Gödel was a Platonist in this sense.

1 Explorers in the realm of numbers

The most fundamental question in the philosophy of mathematics is: what are numbers? The answer given by mathematical Platonism goes roughly as follows. Numbers are objective, real abstract entities that exist in and by themselves, independently of us. Of course, nobody has ever bumped into the number four while crossing the street, and it doesn't make much sense to wonder where in space–time the number four might be, or how much it might weigh, or whether it is red all over. Numbers are immaterial, disembodied entities, just like the famous Platonic ideas: the Idea of Man, the Idea of Beauty, etc. Numbers are intelligible entities, supposedly grasped, so to speak, with the mind's eye, not perceived by means of sensory experience. In the Platonic conception, mathematical sentences have a descriptive content: specifically, they describe these independently existing objects by talking about their eternal properties. Mathematicians do not make up or create mathematical facts, but *discover* them. Mathematicians are the explorers of this intelligible realm and, as G.H. Hardy used to say, to do mathematics is to present the reports of these explorations. That 1,729 is the smallest number expressible in two different ways as the sum of two (positive) cubes would be true even if neither Srinivasa Ramanujan nor anyone else had ever paid attention to this: the number 1,729 is a self-subsistent entity, utterly indifferent to our constructions and theorem-proving activities, completely determined with all its features, including that one. Among the self-declared Platonist mathematicians are Hermite, Hardy, and Roger Penrose, the last of whom I shall talk in the following.

As a philosophy of mathematics, Platonism faces many problems (which does not mean that alternative philosophies of mathematics fare much better). First, there are the epistemic difficulties. Everyone (or nearly everyone) agrees that we can know in a more or less accurate way the physical reality surrounding us by means of our sensory faculties; but many people simply cannot see exactly what this intellectual insight invoked by Platonists is supposed to be, and how it might allow us to grasp purely intelligible and perceptually inaccessible entities. "Mind's eye" is, of course, just a metaphor – but a metaphor of what? Some talk of an "intellectual intuition" ("noetic rays," as David Lewis once lightheartedly said), but it is doubtful that such a thing exists. Kant believed that an intuitive intellect should be an infinite one, i.e., that in order to have a single intellectual intuition, one had to be God.

Second, if Platonism is right then why do mathematicians work so hard to prove things? How come that people just didn't *see* that Fermat's Theorem holds, that is, that $x^n + y^n = z^n$ has no solutions in positive integers when $n > 2$, and we needed some centuries and a proof of more than 100 pages to ascertain the fact? The Platonist claims she has an intuition of what we have called the standard model $\mathbb{N}$ (or of what this model represents), that is, she has some vision of the realm of numbers. According to some, though, this is but a myth: Platonism is a naïve perspective on mathematics; it may have some heuristic value, but it should not be taken too seriously when one engages in philosophical issues concerning mathematics. A famous dictum by Dieudonné, traced back to Bourbaki, has it that professional mathematicians are Platonists on working days, that is, when they do their job, and formalists on Sundays, that is, when they present themselves in the churchyard – and they want to avoid being bothered by the distressing questions asked by philosophers on their metaphysical presuppositions.

2 The essence of a life

Kurt Gödel was a Platonist on Sundays too.

Rebecca Goldstein has claimed that mathematical Platonism was the real essence of Gödel's life. He appears to have embraced a form of mathematical realism from the beginning of his university studies. Given the temperate man he was, though, he kept his convictions hidden for a long time. Gödel's logical formation had taken place in close contact with the famous Vienna Circle, of which his tutor Hans Hahn was an eminent member. And the Vienna Circle treated mathematical statements essentially as "syntax of language," denying them any substantive descriptive content. Platonism in general, with its great bunch of abstract, empirically untreatable entities, was usually considered as a kind of metaphysical obscurantism: as the kind of philosophy from which Moritz Schlick and Rudolf Carnap wanted to free the world once and forever.

Of course, Hilbert's formalist approach could also lead to anti-Platonic positions, since it was based on treating most mathematical sentences (the "infinitary" ones) as devoid of autonomous referential content.[1] The intuitionists led by Brouwer were equally opposed to Platonism in their turn, because of their mathematical constructivism. According to the intuitionists, mathematical entities do not subsist independently of us at all; on the contrary, their properties are determined by mathematical proofs: mathematicians do not discover, but *produce*. Mathematical truths are not true descriptions of non-mental, intelligible, and autonomously existing entities; they are just the theorems proved by mathematicians.

---

[1] The philosophical implications of Hilbert's formalism may actually be more complex. Wittgenstein was a fervent critic of Platonism, and he believed formalism was the other side of the Platonic coin. Although I shall not argue it here, I think he had good reasons for holding this view.

On the whole, the 23-year-old nameless young man who proved the Most Important of All Theorems was surrounded by mathematicians and philosophers who didn't share his ideas on the nature of mathematics at all. So he kept his convictions to himself: he wasn't like them, but they didn't realize it. However, Gödel came to the fore several years later, and did it unequivocally:

> Classes and concepts may, however, also be conceived as real objects, namely classes as "pluralities of things" or as structures consisting of a plurality of things and concepts as the properties and relations of things existing independently of our definitions and constructions.
>
> It seems to me that the assumption of such objects is quite as legitimate as the assumption of physical bodies and there is quite as much reason to believe in their existence.[2]

But what has Gödel's *Theorem* to do with all this? To begin with, Gödel was persuaded that his own mathematical Platonism had been the heuristic guide to the discovery (discovery, not creation) of the Incompleteness Theorem. Second, according to many commentators Gödel took the Theorem itself as a foundation of mathematical Platonism. He had very strong metatheoretical convictions concerning his own work in mathematical logic. According to Rebecca Goldstein, what Gödel aimed at by means of his Theorem was solving a philosophical problem via a mathematical result.

Here is another pronouncement by Gödel, taken from "What is Cantor's Continuum Problem?" – the philosophical essay with which he came to the fore again in 1947:

> The objects of transfinite set theory clearly do not belong to the physical world and even their indirect connection with physical experience is very loose (owing primarily to the fact that set-theoretical concepts play only a minor role in the physical theories of today).
>
> But, despite their remoteness from sense experience, we do have something like a perception also of the objects of set theory, as is seen from the fact that the axioms force themselves upon us as being true. I don't see any reason why we should have less confidence in this kind of perception, i.e., in mathematical intuition, than in sense perception, which induces us to build up physical theories and to expect that future

---

[2] Gödel (1944), pp. 25–6.

sense perceptions will agree with them and, moreover, to believe that a question not decidable now has meaning and may be decided in the future ...

Continued appeals to mathematical intuition are necessary not only for obtaining unambiguous answers to the questions of transfinite set-theory, but also for the solution of the problems of finitary number theory (of the type of Goldbach's conjecture), where the meaningfulness and unambiguity of the concepts entering into them can hardly be doubted. This *follows from the fact that for any axiomatic system there are infinitely many undecidable propositions of this type.*[3]

Goldbach's conjecture is the hypothesis that every even integer (greater than 2) can be expressed as the sum of two primes, and, as Gödel claims, it belongs to "finitary number theory." It has not been proved yet; on the other hand, no counter-example has been provided. Goldbach's conjecture bears this name because it is an old unsolved problem of number theory. However, Gödel insists, we have the clear impression that "Each even integer can be expressed as the sum of two primes" is a perfectly meaningful statement: specifically, according to many it makes sense to suppose that it is true, or false, even though we don't know which of the two is the case. A mathematical Platonist will tend to believe that Bivalence holds for sentences that talk of the realm of numbers: mathematical sentences must be either true or false, provided they are meaningful (and if they aren't, then they are not even real *sentences*). This is the logical-linguistic counterpart of the metaphysical persuasion that the realm of numbers is precisely a realm of self-subsisting, autonomous entities, completely determined independently of our human constructions.

An example I often borrow for my courses of logic is the following. Take the statement: "There are four consecutive 7s in the decimal expansion of π." As we all know, π has an infinite decimal expansion, which means that searching for a proof or a refutation of this claim might spell trouble. However, for the Platonist the statement is in itself true or false. That we don't know which is the right option is an *epistemic* issue: it has to do with what we know or can hope to know about numbers, not with numbers themselves. Like all mathematical entities, π is Out There, perfectly complete with all its properties in the Platonic heaven of numbers. π is utterly indifferent to what we manage to prove

[3]   Gödel (1990), p. 268.

or come to know about it: either those four consecutive sevens occur in its decimal expansion, or they don't – *tertium non datur*.[4]

Now, if we are persuaded that such sentences as "Each even integer can be expressed as the sum of two primes" are in themselves already true or false (that is, such that their negation is true), even though we have a proof neither of the sentence nor of its negation, we will tend to believe that *truth* is distinct from *proof*. And if, as Gödel maintained, mathematical theories describe a fully determined reality, in which each mathematical assertion of a certain kind has to be either true or false, then the fact that a sentence turns out to be undecidable on the basis of the axioms of a theory can only mean that such axioms provide an incomplete description of that reality.

According to the Platonic interpretation, Gödel's Incompleteness Theorem provides a firm foundation precisely for this intuition.

3 "The philosophical prejudices of our times"

It is quite likely that Gödel perceived himself as guided, in his formidable proof of the Incompleteness Theorem, by antecedent Platonic persuasions – by the persuasion that mathematical reality and truth can exceed a full capture by formal systems. Solomon Feferman has reconstructed this story very nicely in "Kurt Gödel: Conviction and Caution," and I shall follow his account.

In 1970 a doctoral student named Y. Balas sent Gödel a letter, asking for explanation on how he had managed to find his undecidable formula. Gödel's answer talks about how he began by searching for a relative consistency proof of analysis in arithmetic, seen as a first step towards the solution of Hilbert's Second Problem. As has been hinted at several pages ago, it is by following this path that Gödel stumbled upon the most unexpected problem for Hilbert's Program. But the letter to Balas was never sent, and remained in the *Nachlass* (luckily, Gödel never threw anything away). In particular, the following passage was crossed out:

[4] By the way, the four 7s are there: they were found by Daniel Shanks and John Wrench Jr in 1962, after developing π for about 5,000 decimals (my students usually smile when I tell them this story).

However in consequence of the philosophical prejudices of our times 1. Nobody was looking for a relative consistency proof because [it] was considered axiomatic that a consistency proof must be finitary in order to make sense 2.A concept of objective mathematical truth as opposed to demonstrability was viewed with greatest suspicion and widely rejected as meaningless.

And "here, in a crossed-out passage in an unsent reply to an unknown graduate student," we have reached "the heart of the matter."[5] We know that Gödel produced a "syntactic" proof, by assuming only the (omega-) consistency of the system he was working on. He did this in order to present his result in such a way as to make it acceptable within the dominant anti-Platonic atmosphere of his time (the "philosophical prejudices"). But his primal intuition concerned mathematical truth, as opposed to provability taken as a syntactic notion, and the metaphysical status of numbers as real entities in the Platonic heaven.

Second, that Gödel believed himself to have obtained by means of his result a refutation of the philosophical positions reducing mathematics to "syntax of language," and mathematical truth to proof, emerges from another great letter that was never sent. The letter dates back to 1974, it was addressed to the sociologist Burke D. Grandjean, and its opening sentence appears as the opening quotation of the second part of this book. Grandjean had sent Gödel a feedback form in order to gather information on our mathematical logician (curious people tend to bother geniuses with lots of questions). Here is a part of Gödel's answer:

> It is true that my interest in the foundations of mathematics was aroused by the "Vienna Circle," but the philosophical consequences of my results, as well as the heuristic principles leading to them, are anything but positivistic or empiricistic … I was a conceptual and mathematical realist since about 1925. I never held the view that mathematics is syntax of language. Rather this view, understood in any reasonable sense, can be *disproved* by my results.[6]

Is this actually the case? Is it the case that, as Gödel appears to have believed, the Incompleteness Theorem, this result in mathematical

---

[5]   Feferman (1983), p. 109.
[6]   Quoted in Goldstein (2005), pp. 112–13.

logic, refutes the idea of mathematics as pure syntax, and validates the metaphysical claim that numbers are real, objective entities in the timeless Platonic sky?

4 From Gödel to Tarski

In the latest edition of their classic handbook *Computability and Logic*, Boolos, Burgess, and Jeffrey affirm:

> Perhaps the most important implication of the incompleteness theorem is what it says about the notions of *truth* (in the standard interpretation) and *provability* (in a formal system): *they are in no sense the same.*[7]

To understand why this is so, let us consider a claim made in the initial paragraph of Gödel's 1931 paper – where, as we know, he was providing the draft of the "semantic" proof:

> From the remark that [the Gödel formula of the system] says about itself that it is not provable it follows at once that [it] is true, for [it] *is* indeed unprovable (being undecidable). Thus, the proposition that is undecidable *in the system* [**TNT**][8] still was decided by metamathematical considerations.[9]

What does this mean? Recall the draft "semantic" proof of the First Theorem that figures in the second chapter of this book, and which followed Gödel's original informal argumentation. One could go down the same path with respect to γ, thus supplying a "semantic proof" of the First Incompleteness Theorem for our Typographical Number Theory. One could reason as follows. After all, we have built within **TNT** a formula, namely γ, which can be read as claiming (via arithmetization) to be unprovable. Now if what it says were the case, then γ would be false. **TNT** would therefore allow us to prove some false statement. Conversely, if we assume **TNT** to be sound or correct

---

[7]   Boolos, Burgess, and Jeffrey (2002), p. 225, italics in the original.

[8]   As usual, Gödel actually referred to (his slightly modified version of) the system of *Principia mathematica*.

[9]   Gödel (1931), p. 19.

(in the usual sense that it proves only truths), then γ must be unprovable in **TNT**. If it actually is unprovable, then γ is just what it claims to be. Therefore, it is true. And so, between γ and ¬γ, γ is the truly true one. Being the negation of a truth, ¬γ is false; therefore, given again the assumption of the correctness of **TNT**, ¬γ in its turn is not provable in **TNT**.

Quite honest and mathematically respectable "semantic" proofs of G1 are available, and they make use of the notion of truth (of the correctness of the system): various proofs of this kind can be found, for instance, in Chapters 3 and 4 of Raymond Smullyan's *Gödel's Incompleteness Theorems* (precisely the fact that they resort to the semantic notion of truth makes them somewhat simpler than the "syntactic" proof we have been dealing with in the first part of this book). In particular, Smullyan's Chapter 4 includes a proof conducted on **TNT**, which assumes the correctness of the theory, and concludes with a true but unprovable formula.[10]

However, whereas the "syntactic" proof of G1 can be formalized within **TNT**, the "semantic" proof cannot.[11] This is because in order to formalize it we would have to express within **TNT** the notion of truth for the **TNT** formulas. But this is ruled out by *Tarski's Theorem* on the *undefinability of truth*, whose proof is closely connected to that of Gödel's Theorem.

Tarski's Theorem concerns more the expressive capacities of the **TNT** language than its theorem-proving power. We already know that the **TNT** formulas are at most denumerable, whereas the properties of natural numbers (the subsets of the set N of naturals) are, by Cantor's Theorem, more than denumerable. So there exist properties of the naturals which cannot be expressed or defined within the language of **TNT**. And being (the code of) a true arithmetic sentence is one of them.

Let us begin by considering what could be taken as a good definition of truth for the (declarative) sentences of *any* given language. To repeat the famous Tarskian example, we can specify a necessary and sufficient condition for the truth of the ordinary English sentence "Snow is white," thus:

---

[10]   See Smullyan (1992), p. 49.
[11]   See ibid., p. 112.

(1)   "Snow is white" is true (in English) if and only if snow is white.

What does (1) claim? This is a biconditional ("if and only if"), whose left side ascribes truth to the sentence at issue. The right side expresses the metalinguistic translation of the sentence itself, that is, its translation in the language in which we specify the truth conditions of the sentence of the object language.[12] Generalizing, we obtain the scheme:

(T) N is true (in such-and-such language) if and only if T,

where N is a name of the sentence to which the feature of being true is ascribed, and T is its translation into the metalanguage. This is the celebrated "T-scheme." Tarski proposed a condition of adequacy (which he dubbed "material") for the definition of truth for a given language by means of his so-called *convention T*: the definition shall be adequate if it is possible to deduce from it all the instances of (T); that is, if for each sentence of the language at issue we shall in principle be able to derive from the definition the corresponding biconditional. Such a characterization would be satisfactory, because the biconditionals of the form (T), taken together, would fix the extension of the truth predicate for the relevant language.

In a formal context in which we want to define truth for the language of **TNT** we need a definition (say in the form of a group of axioms) such that, for any α of the **TNT** language, one can derive the corresponding instance of the T-scheme, that is, something like:

$$(\text{T}_{\text{TNT}})\ T(\ulcorner \alpha \urcorner) \leftrightarrow \alpha,$$

where it turns out that T is the proposed truth predicate for the **TNT** language. Each instance of the T-scheme for the language should be a theorem of the theory within which the notion of truth is defined, that is, it should be possible to obtain it from the bunch of axioms providing the materially adequate definition.

---

[12]   Calling it a "translation" may sound strange, since object language and metalanguage coincide (it's English in both cases). Technically, though, it is a translation. This emerges if we adopt, say, French as the object language, and English as the metalanguage: "La neige est blanche" is true (in French) if and only if snow is white.

Now, suppose such a theory to be **TNT** itself: suppose that the truth predicate for (the language of) **TNT** can be defined, in the Tarskian sense, within **TNT** – that is to say: **TNT** includes its own definition of truth. Then we would have that mixture of language and metalanguage which, as we have seen informally in the first chapter, leads directly to the Liar paradox. Were the notion of truth for **TNT** definable within **TNT**, any instance of the scheme (T_{TNT}) would be a theorem of the system itself. Then we could easily build a formal variant of the Liar paradox within our theory, by exploiting the Diagonal Lemma. We could have:

$$\vdash_{TNT} \lambda \leftrightarrow \neg T(\ulcorner \lambda \urcorner).$$

Now λ is a formal Liar: a fixed point of the property of being not true, i.e., a sentence which claims of itself to be untrue. So we could build in our Typographical Number Theory the Liar paradox by means of the corresponding instance of the T-scheme for λ (such an instance would be a theorem of the system); and we could obtain $T(\ulcorner \lambda \urcorner) \leftrightarrow \neg T(\ulcorner \lambda \urcorner)$ (the Liar is true iff it is untrue). So Tarski's Theorem concludes:

(TT)   If **TNT** is consistent, then it is not possible to define the notion of truth for the sentences of **TNT** within **TNT**.[13]

We know that the Gödel sentence γ of **TNT** leads us close to a paradox, but in the end escapes it; this is because Gödel replaced the notion of truth with that of provability. γ produces no paradox, whereas λ does (or would, were truth definable). So it is impossible to define (*a fortiori*, to recursively enumerate) the set of truths of **TNT**, on pain of contradiction. But the set of theorems of **TNT**, on the contrary, *is* recursively enumerable, and therefore (weakly) representable in **TNT**; and here emerges the discrepancy between provability and truth we were looking for.

Gödel wrote before the publication of the Tarskian result, but he had widely anticipated it. In Feferman's words:

> If truth for number theory *were* definable within itself, one could find a precise version of the Liar statement, giving a contradiction. It follows that truth is not so definable. But provability in the system *is* definable,

[13]   See Smullyan (1992), pp. 26–7.

so the notions of provability and truth must be distinct. In particular, if all provable sentences are true, there must be true non-provable sentences ... We may conclude from all this that the concept of truth in arithmetic was for Gödel a definite objective notion, and that he had arrived at the undefinability of that notion in arithmetic by 1931.[14]

Godel's claim that the undecidable sentence of the formal system "still was decided by metamathematical considerations" can be translated, after Tarski, by saying that γ, unprovable within **TNT**, is nevertheless provable in a metatheory capable of dealing with the semantics via the notion of truth – a notion indefinable in **TNT** because of (TT). Roughly, the definition of truth for the language of **TNT** "can be given in an extension of that language to a next higher type,"[15] and here we can carry out the "semantic" reasoning leading us to the proof (of the truth) of γ, that is, to deciding γ.

We are now in a position to fully appreciate the fundamental sense of endnote 48a of the 1931 paper, where Gödel tells us something about the roots of the Incompleteness Theorem:

> The true reason for the incompleteness inherent in all formal systems of mathematics is that the formation of ever higher types can be continued into the transfinite ... while in any formal system at most denumerably many of them are available. For it can be shown that the undecidable propositions constructed here become decidable whenever appropriate higher types are added.[16]

5   Human, too human

The "semantic" version of (the proof of) G1 tells us that, if **TNT** is sound, then it doesn't prove a sentence which is nevertheless true. Generalizing: any formal system sufficiently strong, etc., etc., which proves *only* arithmetical truths, cannot prove *all* arithmetical truths. And this is how truth seems to transcend provability and to corroborate mathematical

---

[14]   Feferman (1983), p. 107.
[15]   Ibid., p. 109.
[16]   Gödel (1931), p. 44.

Platonism. Formal systems turn out to be a human – too human – construction. That truth surpasses proof means that mathematical reality transcends the finitary – that is, too human – requirements dominating the construction of formal systems. Here is a summary of this Platonic interpretation of G1:

> [γ] is both unprovable and, given that *that's what it says*, it's also true. We haven't shown that it's true by finding a proof for it within the formal system, using the purely mechanical rules of that system, that is, by deducing it. Rather, we've shown it's true by, ironically, going outside the system[17] and showing that no proof for it *can* be produced within the formal system … Gödel's result, in effect, proclaims the robustness of the mathematical notion of infinity; it can't be drained of its vitality and turned into a ghostly Kantian-type idea hovering somewhere over, but without entering into, mathematics. The mathematician's intuitions of infinity – in particular, the infinite structure that is the natural numbers – can no more be reduced to finitary formal systems than they can be expunged from mathematics … There remains something – always – that eludes capture in a formal system. It was in this metalight that Gödel viewed his incompleteness theorems.[18]

But is it really so that the mathematical result the Theorem consists in substantiates a conclusion of such philosophical importance? The "semantic" proof of G1 actually is the proof of a conditional: what we are told is that *if* **TNT** is sound, then some truth (that of γ) escapes the system, so **TNT** is semantically incomplete. But how can we know for sure *that* γ, besides being unprovable, is true? In order to obtain this conclusion, one would need a proof of the soundness of **TNT**, that is, of the antecedent of that conditional. We have good evidence that **TNT** is sound indeed; but that the same can be said of any formal system falling under some version or other of Gödel's Theorem is far from obvious. I shall not enter into this here, for the topic is to be discussed extensively in the next two chapters: it is at the core both of the skeptical interpretations of the Theorem, and of the debate on the (in)famous Gödelian arguments in the philosophy of mind. For the time being, we shall be content with the following remark: the anti-Platonist can stress that, even though a "semantic" version of G1 is acceptable, what

---

[17] I'll come back to the expression "going outside the system" in the following.
[18] Goldstein (2005), pp. 183 and 186-7.

has been established is just the conditional claim that if the system is sound, then its undecidable Gödel sentence is true.

Now the Platonist can reply that there is no need to ascertain that γ is true in order to obtain the semantic incompleteness of **TNT**, and can account for this by means of an argument which has already been hinted at towards the end of the first part of the book. However, precisely such argument sheds light on an assumption in which the Platonist's Platonism is already embodied: the Principle of Bivalence (for mathematical statements). Recall that the syntactic incompleteness of **TNT** produces a semantic incompleteness if we accept that Bivalence holds for the **TNT** formulas. In this case, there is no need to argue on the issue of the truth of γ. The point now is that γ is either true or false, that is, either γ is true, or ¬γ is. But **TNT** doesn't prove either, as the "syntactic" proof based only upon the (omega-) consistency of the system suffices to testify. So if **TNT** is (omega-) consistent, then there exists in any case some truth which the system cannot prove. Generalizing: if any sufficiently strong (omega-) consistent formal system, as Gödel has told us, includes undecidable sentences of finitary number theory, and these sentences must be either true or false, then the (First) Incompleteness Theorem, in its *syntactic* version, establishes in a precise way that mathematical truth surpasses what can be proved in any single such system.

Now a philosopher of mathematics of formalist or in general anti-Platonic attitude may also contest the conclusion without taking into account the further development according to which γ is the true one (true in the standard model ℕ, that is). In Gabriele Lolli's words:

> Behind the vulgar formulation [of the Theorem, i.e., the one to the effect that there are *true* unprovable sentences] hides the archaic idea that the structure ℕ of natural numbers is clear and distinct, that the arithmetical theorems are the sentences which are true in it, and that everyone intuitively knows what it means to say that an arithmetical sentence is true. Nothing is more distant from the conclusion reached in the eighteenth century, when the conviction that the theorems are truths concerning a precise reality, known to the mathematician, faded ... The incompleteness theorem does not presuppose the existence of the structure ℕ, but only that the theory is consistent. The theorem does not prove that one of the sentences is true, but only that if the theory is consistent then both are unprovable.
>
> If the structure ℕ existed, and were clear to the mind, then the issue of consistency would be meaningless, just as it would be meaningless to

posit the consistency of the theory as a hypothesis, for if its axioms are true in ℕ then it is consistent for sure.

We certainly use the notion of truth in a legitimate and innocuous way at times … But if instead of thinking about the intuitive notion of truth, one wants a more precise and sophisticated one, then Tarski's theorem on the arithmetical undefinability of truth enters the stage. The theorem does not mean that truth is absolutely ineffable – we are talking about it, after all. But it testifies that its definition is not arithmetical, it requires a richer language and other (disputable) assumptions. As a matter of fact, it has been clarified that the definition can be given in second-order logic … Truths differ from theorems, then, not because there are fewer theorems than truths, but because talking about truth in an infinite structure is much more cumbersome and less intuitive than talking about proof.[19]

Shall we limit ourselves, then, to a merely "syntactic" declaration, that is, that neither γ nor ¬γ is a theorem of **TNT**, if **TNT** is (omega-) consistent? Is the subsequent claim that, given any formal system fulfilling the conditions of applicability of the Theorem, there exists some truth which is unprovable in it, a merely philosophical and non-mathematical tenet? Some have claimed that this is the case, for instance – albeit in a different context – Juliet Floyd and Hilary Putnam:

> That the Gödel theorem *shows* that (1) there is a well-defined notion of "mathematical truth" applicable to every formula of PM; and (2) that, if PM is consistent, then some "mathematical truths" in *that* sense are undecidable in PM, is *not* a mathematical result but a metaphysical claim.[20]

If this is the case, then the Platonic reading of the Incompleteness Theorem may not get off the ground. Contrary to what Gödel himself seems to have believed, there is no philosophical result *based upon* a mathematical theorem here.

Of course, the anti-Platonist, too, is talking philosophically, not mathematically. The key issue, in fact, has to do with the existence and nature of the mythical standard model ℕ; that is to say, of the intuitive interpretation of our Typographical Number Theory; and that is to say, of the realm of natural numbers our famous teacher told us about. It is

---

because people have *already* made up their minds on their metaphysical preferences (for or against Bivalence for arithmetical statements; for or against the idea that the world of numbers is completely determined no matter what people manage to prove or disprove) that people argue and disagree on what the Incompleteness Theorem should teach us on this. The Platonist maintains she has an intellectual intuition of this infinite mathematical reality. She cheerfully concedes that this intuition is not a complete and perfect grasp; after all, we have an extremely approximate knowledge of the external physical world; however, only sophisticated skeptical arguments give rise to qualms about the existence and actuality of this physical reality surrounding us. By granting that we have an imperfect knowledge of the standard model, the Platonist believes she can also account for the need for proofs in mathematics: we need proofs because in many or most cases our picture of numbers is not clear at all, and proof has to replace intuition.

The anti-Platonist replies that the Platonist can at most *believe* she sees something, but there actually is nothing to see, for the mythical $\mathbb{N}$ is a myth. Those who speak of true arithmetical statements in the Platonic sense presuppose that $\mathbb{N}$ is accessible to knowledge like the Euclidean space, for they tell us nothing about how one establishes that such statements are true. This is "not so different from someone who has been taught since he was a child to believe he has direct knowledge of some things concerning God - things which have actually been heard in the adults' discourses"; assuming one has this kind of knowledge, then "one convinces himself that human beings possess some faculty capable of providing such a knowledge," that is, "in one case, faith; in the other, a special intuition."[21]

Overall, one party makes a claim on the basis of an allegedly fundamental and intuitive knowledge; the other party denies such intuition exists. We are in the middle of a violent clash of ... intuitions. And it seems, as a Wittgensteinian might say, that the chain of reason ends here.

---

[21] Lolli (2004), pp. 166–7. "It is false that people learn mathematics by contemplating structures objectively offered to their eyes. Those who claim that mathematicians contemplate pre-existing structures with the mind's eye have simply forgotten their first contact and their complex familiarization with mathematics ... The Platonic account is false and misleading" (Lolli (2002), pp. 104–5).

10

Mathematical Faith

A maxim attributed to the great mathematician André Weil has it that God exists since mathematics is consistent, and the Devil exists since we cannot prove it. Gödel's Second Theorem according to many has such religious consequences, and in this chapter we will ponder whether and in what sense this is the case. We shall investigate the mathematical nature of consistency proofs, their philosophical import, and the possibility that G2 undermines either. We won't need to bother God or the Devil, but we will have to talk about faith – at least, about mathematical faith. We will also talk about what is usually opposed to faith, that is, skepticism. As we will see, though, skepticism and faith can support each other, just as some officially opposed political parties do when they see a yet more dangerous rival on the horizon. Needless to say, in this case the rival is philosophy or, better, a certain philosophy.

Sextus Empiricus, the most well-known skeptic of antiquity, believed skepticism to have a therapeutic function. The need for a philosophical *ubi consistam* was for him a kind of pathology, a mental disorder. The skeptic usually starts as a philosophical addict looking for ultimate truths. Frustrated by the battalions of arguments and counterarguments used by opposing philosophers about what such truths would be, the skeptic falls, exhausted, into a state of suspended judgment: he doesn't know who to agree with anymore. And precisely this state brings unexpected relief, by setting him free from the need for certainty. Sextus refers to the legend of the painter Apelles, who was trying to paint some foam on the mouth of a horse. After numerous fruitless attempts, Apelles threw the sponge – that is, he actually threw the sponge he used to clean the mistakes – against the picture. By complete chance, the sponge hit the horse's mouth and produced a perfect painted foam.

According to some mathematical skeptics, one should tell a similar story with respect to Gödel's Theorem. Mathematical logicians with a passion for philosophical foundationalism begin to search for a proof of the consistency of arithmetic. Various attempts fail, until one of them produces a completely unexpected result: it turns out that the consistency of arithmetic cannot be proved. Precisely this negative result, though, provides the emancipation people were looking for: now mathematicians can work without philosophers bothering them with foundational requests, for it has been proved that there can be no such foundations.

Is this actually the case? Let us investigate.

1 "I'm not crazy!"

We know that Hilbert took a stance on the crisis in the foundation of mathematics in order to provide a secure ground for analysis, and especially arithmetic, after the emergence of the set-theoretic paradoxes. In his Program he proposed to build a metamathematical finitary consistency proof for formalized arithmetic. The problem of supplying a consistency proof is particularly pressing for a formalist, more than for a Platonist. If such a theory as our Typographical Number Theory has a model in the mythical $\mathbb{N}$, which the Platonist claims to "see" more or less clearly, then **TNT** is certainly consistent: all its axioms hold in that model. A formalist, however, cannot allow herself such "semantic" reasoning, referring to models and truth: she needs a purely syntactic consistency certificate. Conversely, Hilbert believed that the consistency of a formal system was *sufficient* to guarantee the existence of a model. Gödel was opposed to this persuasion (and when he gave the first exposition of his Theorem in Königsberg, he began precisely by criticizing it).

As we have seen, it is the Second Incompleteness Theorem that spells trouble for Hilbert's Program. G2, as formulated in the first part of our book, tells us that if Typographical Number Theory is consistent, then it cannot prove its consistency statement **Cons$_{TNT}$**. But G2 can be generalized, as usual, to any consistent and sufficiently strong formal system. It applies to the current mainstream set theories, and particularly to **ZFC**, the Zermelo–Fraenkel set theory with the Axiom of Choice, which, as

I have said, is something like the regular set-theoretic formal system among mathematicians. Whereas in **TNT** we can formalize the arithmetic reasoning usually qualified as elementary, **ZFC** is a particularly powerful system within which one can prove more or less all the theorems of ordinary mathematics. In particular, there are no *mathematical* conjectures or problems which have ever been dealt with by *mathematicians* which are provably undecidable in **ZFC**.[1] One should keep this difference between **TNT** and **ZFC** in mind because, even though philosophical discussions on the Incompleteness Theorem usually overlook it, we shall see that it has some impact on the problem of consistency proofs.

G2 is at the core of the *skeptical interpretations* of Gödel's results. The skepticism at issue is, again, a kind of anti-foundationalism. It is claimed that the impossibility of proving the consistency of a consistent formal system within the system itself entails some unavoidable uncertainty for the working mathematician: we can never be sure that the formal systems we work with are consistent. Some even claim that, because of the Second Incompleteness Theorem, "to exaggerate perhaps only a little, the consistency of systems like Peano (even Robinson) arithmetic must be taken in faith."[2] Certainty and faith are epistemic facts, and the skeptical interpretations of G2 are declarations of epistemic desperation.

Here is a characterization of the skeptical interpretation provided by Michael Detlefsen:

> G2 somehow shows that any consistency proof for a theory *T* of which it (G2) holds will have to make use of a premise-set that is more dubitable than the premise-set of any proof constructible in *T*.[3]

---

[1]   Whereas there are *set-theoretic* conjectures which are provably undecidable within **ZFC**. The most celebrated among them is the *Continuum Hypothesis*, corresponding to the first of Hilbert's Problems. The Continuum Hypothesis is the thesis that there are no infinite sets whose size is intermediate between that of the naturals and that of the real numbers. Gödel contributed to the solution of the problem in the 1940s, by showing that the negation of the Hypothesis is not provable from the axioms of standard set theory. In 1963 Paul Cohen, at the time a young mathematician at Stanford, showed that the Hypothesis is actually independent from the axioms of set theory, that is, it is undecidable with respect to them, by using a mathematical technique called *forcing* – and he was the first (and, so far, the only) logician to thereby earn the Fields medal, that is, the Oscar of mathematics.

[2]   Dunn (1986), p. 161.

[3]   Detlefsen (1979), p. 131.

Skeptical readings of the Second Theorem have been advanced by various authors, including Michael Resnik, Paul Cohen, E.W. Beth, and even Nagel and Newman:

> [Gödel] proved that it is impossible to establish the internal logical consistency of a very large class of deductive systems – number theory, for example – unless one adopts principles of reasoning so complex that their internal consistency is as open to doubt as that of the systems themselves.[4]

To assess the force of these claims, as is often the case in philosophy, we need to clarify the meanings of some words. Specifically, the problem is that the meaning of "doubt" is doubtful. As Smullyan has observed, any formal system **S** proving its own consistency, in a sense, could not solve any doubt whatsoever on the consistency of **S**; on the contrary, it would *raise* some serious ones.

Suppose someone you know begins to declare to everyone, again and again, that she is not crazy, or that she never lies. After a while, most people will start to think that the person lies, or that she is a bit crazy. One may produce an *excusatio non petita* once in a while. But when one begins to systematically declare that she is normal – she never lies, she isn't crazy – there's something abnormal going on. Ordinary people typically say nothing of the sort.

Now if a given formal system **S** of the relevant kind is inconsistent, and its underlying logic is classical, then **S** *does prove* its own consistency, in the sense of proving its consistency statement (**Cons$_S$** say), for the plain reason that, via Scotus' Law, it can prove anything. To believe in the consistency of a formal system **S** on the basis of its proving **Cons$_S$** is like believing that someone is not crazy on the basis of her claiming that she isn't – which is precisely what many unfortunate persons secluded in psychiatric hospitals claim. "The fact that P.A. [Peano Arithmetic, that is, **TNT**], if consistent, cannot prove its own consistency – this fact does not constitute the slightest rational grounds for doubting the consistency of P.A."[5]

[4]   Nagel and Newman (1958), p. 5.
[5]   Smullyan (1992), p. 109.

2   Qualified doubts

Precisely owing to these considerations, on the other hand, what Gödelian skeptics have in mind cannot be such a stupid thing! Let us look at a qualified version of the skepticism at issue:

> Gödel's Second Theorem implies that the consistency of *Principia* can be mathematically proven only by conjecturally assuming the consistency of *Principia* outright (which is what mathematicians implicitly do in practice), or by reducing the consistency of *Principia* to that of a stronger system, thereby beginning an infinite regress.[6]

Now this doubts story is more precise, for the notion of doubt can be specified by resorting to that of a *stronger* formal system, and this notion is somehow more precise. As you may remember from the first part of the book, according to some standard definitions, given two formal systems S_1 and S_2, one claims that S_2 is stronger than S_1 if and only if the former proves everything the latter does, and also proves something more (that is, the set of theorems of S_1 is a proper subset of the set of theorems of S_2).

However, when philosophers talk about some formal system being stronger than some other, they often have in mind different – and more substantive – issues. For instance, if some formal system extends some other by means of the addition of an insignificant axiom, philosophers will be reluctant to say that the former is "stronger" in any but a merely technical sense. Conversely, if a system S_1 doesn't prove some of the things proved by a system S_2, but includes a "strong" axiom in an informal but pregnant sense, which S_2 lacks, philosophers will tend to consider it stronger. We shall soon see that distinguishing different senses of "strong" in this context is quite useful in some cases.

Next, we should bear in mind that a consistency proof is a proof of the fact that a given formal system has a quite weak syntactic property. We already know that soundness, that is, the semantic property of proving just truths (or, if you prefer, things that are true in the standard or intended model), entails consistency; but the converse does not hold. Consequently, proving that a given formal system is consistent,

---

[6]   Kadvany (1989), p. 165.

that is, that it doesn't contradict itself, does not rule out the case that the system proves lots of falsities – just as a consistent liar, it has been remarked, can make false claims without self-contradicting. A formal system proving falsities is not exactly the kind of thing we would resort to when doing mathematical logic.

There's a particularly striking example of this situation. Take our Typographical Number Theory again. G2 tells us that, if **TNT** is consistent, then it cannot prove its consistency statement, **Cons**$_{TNT}$. We also know that **Cons**$_{TNT}$ is equivalent to the Gödel sentence γ, and this fact in its turn can be proved in **TNT**. **Cons**$_{TNT}$ is a case of an undecidable **TNT** sentence. Now suppose we add to the system, as an axiom, the *negation* of its consistency statement, thereby obtaining a system, say, **TNT**$_{NonC}$ = **TNT** + ¬**Cons**$_{TNT}$. Given that **Cons**$_{TNT}$ is undecidable within **TNT**, **TNT**$_{NonC}$ is consistent, provided that **TNT** is. So we have that **TNT**$_{NonC}$ is a consistent theory including a false sentence: it is a consistent theory which "claims" (proves) "**TNT** is inconsistent"; that is, it declares, falsely, the inconsistency of Typographical Number Theory and, by entailment, of *itself*. Gödel's Theorem, as Franzén has pointed out, in this case does not raise doubts, but solves them. One might have qualms about the consistency of **TNT**$_{NonC}$, but the incompleteness result entails that the consistency of the theory can be established as a consequence of the consistency of **TNT** itself.[7]

Here is another hint I borrow from Franzén, which shows how issues of consistency can quickly become quite subtle. We know that **TNT** is not finitely axiomatized, because of the induction scheme. However, given any *finite* subset K of axioms of **TNT**, the consistency of K *is* provable within **TNT**. Now a formal proof in **TNT** will always be, by definition, a finite sequence of formulas; it is therefore bound to exploit at most a finite number of axioms. So if any subset of axioms of **TNT** is consistent, then all the proofs carried out in the system are consistent. Thus, if any finite subset of axioms of **TNT** is consistent, **TNT** is consistent. The whole argument can be formalized within **TNT** in its turn, being a simple reasoning involving only elementary principles.[8]

However, even though **TNT** can prove, for any finite set K of axioms of **TNT** (the formalization of) "The theory with all the axioms in K is consistent," **TNT** cannot prove (the formalization of) "Given any finite

7   See Franzén (2005), p. 111.
8   See ibid., p. 103.

set K of axioms, the theory with all the axioms in K is consistent." **TNT**, in the end, seems to get *quite close* to a proof of its own consistency.[9]

3   From Gentzen to the *Dialectica* interpretation

Hilbert knew well that a *non*-finitary consistency proof for arithmetic would have constituted an insufficient reply to the skeptics, for the simple reason that the skeptics of his time raised doubts precisely on the reliability of infinitary reasoning. What he aimed at with his underspecified finitism, required for the metamathematical consistency proof for formalized arithmetic, was an unassailable epistemic base. We know that G2 takes the stage precisely here. Hilbert's Program was based on the idea of proving the consistency of formal systems somewhat stronger than our Typographical Number Theory, such as **ZFC**, by means of strictly finitary methods. But G2 tells us that we cannot prove the consistency even of **TNT** by using the kind of methods Hilbert theorized, since they would certainly be formalizable within **TNT**.

However, as has been revealed in the first part of the book, Gödel did *not* believe he had knocked out the Program because of G2. In the 1931 paper he was quite explicit on this:

> I wish to note expressly that Theorem XI … do[es] not contradict Hilbert's formalistic viewpoint. For this viewpoint presupposes only the existence of a consistency proof in which nothing but finitary means of proof is used, and it is conceivable that there exist finitary proofs that cannot be expressed in the formalism [of the system].[10]

One should say at the outset that things went more or less as Gödel had foreseen (and in this case as in many others, Gödel's Theorem, which is pigeonholed as a "limitative" theorem, or a "proof of impossibility," actually opened up new perspectives and horizons by pushing people

---

[9]   On these issues Feferman (1960), whom I have already quoted, has had a deep impact. What is at issue here is precisely the capacity of some formal systems, called *reflexive*, to prove the consistency of their finite subsystems.

[10]   Gödel (1931), p. 40. As usual, Gödel referred to the system **P** he was working on.

into the search for new mathematical techniques). In 1935 the mathematician Gerhard Gentzen, coming from Hilbert's school, obtained a consistency proof for **TNT** based on a special kind of extended mathematical induction, called "transfinite induction." The proof was published one year later. Gentzen successfully exploited an induction on a segment of the ordinals up to the Cantor transfinite ordinal called "ε_0" (on which I will say something in the following chapter). We are not interested in the technical details of the proof, though; the point of the procedure is that the induction doing the job concerns an ordering which is, so to speak, much "longer" than that of the series of the naturals.

Gentzen's proof provides a good case for distinguishing, as has been flagged above, different senses for "stronger." Gentzen's proof is not a proof in elementary arithmetic, formalizable within **TNT**; in this sense, it actually involves "stronger" principles than those **TNT** can account for. But in another sense, things are not so simple, for the induction at issue is applied only to a very limited set of arithmetic properties. Therefore, this consistency proof "extended [**TNT**] in one direction, but restricted it in another." And "Gentzen's proof, along with the whole subject of 'ordinal analysis' to which it gave rise, is very technical and stands in no simple relation to any doubts that people may have about the consistency of [**TNT**]."[11] To put it otherwise, establishing whether proofs of this kind provide a successful reply to skeptics is not so simple; such a reply would have to resort to some *philosophical* line of argumentation anyway.

Gödel directly worked on consistency proofs for arithmetic. He did this in an important paper published in 1958 (but including results he had obtained years before) in the analytic review *Dialectica*. As a matter of fact, this article – the last paper published by Gödel during his life – included first of all a quantifier-free proof interpretation of intuitionist arithmetic, which is why the theory advanced by Gödel in the text has been called "the *Dialectica* interpretation." Gödel drafted a theory of computable functions of finite type, and proposed to apply it to prove the consistency of formalized arithmetic and also of analysis. Gödel's idea was that "it is practically certain that concrete finitary methods are insufficient to prove the consistency of elementary number theory, and

---

[11] Franzén (2005), p. 107.

some abstract concepts must be used in addition."[12] Hilbert's Program, then, was not left to die, but brought back in a more liberal version.

4   Mathematicians are people of faith

Franzén also points out that simpler and mathematically more accessible consistency proofs are on the market. In such proofs, more than the consistency of **TNT** is proved, namely, its soundness: that all the arithmetical theorems of **TNT** are true. The sample proof comes in three steps: (1) one defines the notion of arithmetical truth; (2) one proves that, given the definition, all the axioms of **TNT** are arithmetical truths; and (3) one proves that the rules of inference of the system are reliable, that is, truth-preserving. The reasoning involved in such a proof is of just the kind that can be formalized within **TNT**, but with the addition of bits of set theory. Suppose the set-theoretic machinery is provided by **ZFC**. This is *much* more than we need, as Franzén points out (second-order Peano Arithmetic would also do); he mentions a subsystem of **ZFC**, called **ACA**. But since **ACA** is little known, whereas **ZFC** is the mainstream set theory among mathematicians, let us stick to the latter. Now the problem is that the skeptical doubts on consistency can easily be transferred to the set-theoretic framework itself: **ZFC** is stronger than **TNT** (in any sense of "stronger"), and Gödel's Theorem, of course, applies to **ZFC** too.

Franzén's reply? This is essentially based on the fact that proving the consistency of **TNT** in standard set theory (**ZFC**, or some restriction) is ordinary mathematical practice, and no mathematician would question the acceptability of the proof:

> In particular, an insistence on a hypothetical interpretation of an arithmetical theorem ... is rarely heard from mathematicians or anybody else in connection with proofs of ordinary arithmetical statements. For example, one never encounters as a response to the claim "Andrew Wiles proved Fermat's last Theorem" the objection "No, no, that's not possible – all he proved was that if ZFC is consistent, then there is no solution in positive integers to $x^n + y^n = z^n$ for $n > 2$" ... Gödel's theorem tells us nothing

---

[12]   Kleene (1986), p. 139.

about what is or is not doubtful in mathematics. To speak of the consistency of arithmetic as something that cannot be proved makes sense only given a skeptical attitude towards ordinary mathematics in general.[13]

A philosopher of mathematics with skeptical inclinations may reply, to begin with, that there is a specific problem concerning the consistency of **ZFC**. We can prove the arithmetic soundness (therefore, consistency) of **TNT** by resorting to the ordinary set-theoretic machinery – to bits of it, indeed. So the proof can legitimately be qualified as "ordinary mathematics." But **ZFC** is a *background* theory, for there is a large consensus on the fact that within **ZFC** one can formalize, as has been said, essentially all the proofs of ordinary mathematics, elementary or not. Contraposing: if a problem is proved unsolvable in **ZFC** (as happened with the aforementioned Continuum Hypothesis), mathematicians will have a strong tendency to quit taking it as a mathematical problem. But Gödel's Second Theorem applies to **ZFC** as well. A consistency proof for **ZFC**, thus, given G2 cannot be formalized within **ZFC**. Therefore, such a consistency proof, if available, *would not be ordinary mathematics*. That **ZFC** cannot decide its own consistency statement should *not* be seen as an ordinary mathematical problem. And such a conclusion would be established by resorting, among other things, to the Second Theorem.

In fact, on the one hand there are no consistency proofs for set theory comparable to Gentzen's for formalized arithmetic. On the other hand, the consistency of **ZFC** can be proved by assuming strong axioms of infinity, which do not belong to ordinary mathematics. So as Franzén admits, "it is a perfectly reasonable observation that these consistency proofs for ZFC are not ordinary mathematical proofs and that such proofs do not establish that the consistency of ZFC can be proved in the same sense as Fermat's Last Theorem has been proved."[14] Besides, these very axioms are not intuitively transparent at all (unless you are Gödel, perhaps), and raise significant epistemic problems in their turn. Hence, one understands Michael Resnik's conclusion:

> What about the consistency of all mathematics or of some strong system for set theory? How do we answer the sceptic? Since here a convincing

---

[13] Franzén (2005), pp. 111–12.
[14] Ibid., p. 108.

proof is not possible, we have established that the sceptic demands too much. We cannot be certain that our axioms are free from contradiction and must treat them as hypotheses which may be abandoned or modified in the face of further mathematical experience.[15]

Second, our Sextus Empiricus of mathematics may add that mathematicians are perfectly right in relying on their background theory – for how could they carry out their work otherwise? This cannot be a confirmation of the theories at issue in any sense, of course (ordinary engineers can ordinarily rely on Newtonian mechanics as their background theory, but Newtonian mechanics is, in Popperian terms, falsified). More importantly, this mathematical reliance might be nothing but faith, given a suitably restrictive definition of "faith." The problem with consistency proofs has nothing to do with their mathematical adequacy; it concerns their philosophical role when one faces claims of epistemic skepticism. Skeptical doubts change nothing from the point of view of working mathematicians. Their healthy function consists in pensioning off certain foundationalist ambitions – such as those embodied in Russell and Frege's logicism, or in Hilbert's Program. As for the latter, there is widespread agreement on the fact that G2 *does* have a deep impact on it, at least in the sense that it rules out, as we have seen many times, a consistency proof for **TNT** which conforms to Hilbert's severe restrictions. And some salute this as a salutary fact; among them, Wittgenstein, who was not a skeptic in the sense now under consideration, but who always spoke against the search for consistency proofs, considered a dangerous philosophical intrusion in the mathematician's work:

> And if they now demand a proof of consistency, because otherwise they would be in danger of falling into the bog at every step – what are they demanding? Well, they are demanding a kind of *order*. But was there *no* order before? – Well, they are asking for an order which appeases them now. – But are they like small children, that merely have to be lulled asleep?[16]

On the other hand, the skeptical attitude in its general form actually goes beyond what Gödel's Theorems have to say, in mathematics as

---

[15]   Resnik (1974), p. 128.
[16]   Wittgenstein (1956), p. 101e. I'll come back to Wittgenstein's stance on the issue in the final chapter.

well as elsewhere. The skeptic tries to entrap us, by first asking for a consistency proof for a formal system on whose consistency there exist "doubts," and then raising further doubts on the demonstrative principles adopted in the proof itself. To this kind of strategy Aristotle provided a reply some millennia ago, when he remarked in the *Metaphysics*:

> Not to know of what things one should demand demonstration, and of what one should not, argues want of education. For it is impossible that there should be demonstration of absolutely everything (there would be an infinite regress, so that there would still be no demonstration).[17]

If we have qualms about the consistency of a formal system for arithmetic – say **TNT** – we can prove the consistency of the system in another theory – say **ZFC**. Now if we claim that we need a consistency proof for **ZFC**, we find ourselves involved in a regress. If we want to stop somewhere, we have to do it by justifying some principle without resorting to further *proofs*, but by way of an appeal to such things as intuitiveness, utility, evidence, whatever these notions mean for philosophers. And if the skeptic now attacks *these* notions, Gödel's Theorem remains silent on the issue.

---

[17]   *Metaphysics* 1006a 7-9.

11

Mind versus Computer: Gödel and Artificial Intelligence

Human beings sometimes want to compete with the computers they have created in some intellectual activity or other, as when we have some world champion of chess play against a chess software (Kasparov versus *Deep Blue*). As the speed and power of computing machines increases, some people are reassured by the fact that the human player can still prevail over the machine, and feel some sort of discomfort when it's the machine that triumphs. One does not need to fear, once machines far too clever and powerful for us have been produced, that the creature may turn against the creator, like Hal9000 in *2001: A Space Odyssey*, or the robots of the *Terminator* series. In fact, many of us are bothered by the mere idea that a computer could be smarter than us.

And in fact, according to some, Gödel's theorem has something to say even on this.

1 Is mind (just) a program?

We all know that today's computers outstrip us in performing calculations. However, some people think that there is something special about human intelligence: that this special ability does not consist merely in carrying out computations (that is, in applying effective, mechanical procedures, therefore algorithms); and that its only owners are the special carbon-based beings supplied with a brain (or maybe with an immortal soul) that we are, so that it cannot be taken over by robots, electronic circuits, and other silicon-based impostors.

This view is not shared by the supporters of artificial intelligence (AI), in the version John Searle called *strong*. Strong AI fans believe the fuzzy set of cognitive abilities we label "intelligence" to be realizable by computer programs. They believe, that is, that to think *is* to compute, i.e., to carry out information processing, operating on data by means of effective, algorithmic rules – which is exactly what computers do. It is conceded that an algorithm capable of capturing the whole activity of a human mind would be very different from a Microsoft Office Professional suite. However complex, though, it would still be an algorithm, so it could in principle be implemented on a suitably fast and powerful computer.

Now, after what we have learnt in the first part of the book, it is not too difficult to see the connection between formal systems, that is, what Gödel's Theorem in its classic formulations officially deals with, and computers. We know that the formal systems to which the Theorem applies are those (omega-) consistent systems capable of representing the recursive functions, and we also know that the equivalence between recursive and Turing-computable functions has been established. Now a universal Turing machine is an ideal machine of which computers are physical realizations.[1] All modern computers are, or embody, something corresponding to a universal Turing machine, the differences between them consisting only in the amount of memory they are supplied with, and in how fast they can be in processing their data. And the equivalence between formal systems and computers or Turing machines follows from the fact that for any formal system one can program a computer so as to mechanically produce all and only the theorems of the system; conversely, for any way in which one can program a computer to work as an automatic theorem-proving machine, there exists some formal system having as theorems precisely the formulas produced by the computer.

Some philosophers and mathematicians have attempted to obtain from the Gödelian result an argument to the effect that the human mind is, if not superior, at the very least irreducible to any computing machine, in a sense we shall explore soon. There are different such arguments around: people speak specifically of "Gödelian arguments"

---

[1] With the limitation you may remember: a Turing machine has an infinite tape, so it is "idealized" at least in this respect, whereas the memory of a real computer, however big, is always finite.

in artificial intelligence. One can find them phrased directly in terms of Turing machines or computers, or else in terms of formal systems; and they have originated a huge debate in the philosophy of mind.The literature on the subject is far too vast to be taken into account in this book, so I'll have to limit myself to a few remarks.

Two things at the outset. First, it is fair to say that the Gödelian arguments are widely criticized, and nowadays few scholars think that they are conclusive against the basic tenets of AI. Second, what *I* think of this story is (a) that, despite being actually mistaken, they are so in an interesting way, and (b) that there exists a kind of variant of these arguments which can be put to good work. However, rather than legitimating claims about our superiority or difference with respect to the machines, it tells us something about what *we can know about* this. And by doing so, it also tells us something important about our mind, about what we can understand of ourselves, and maybe even about the nature of mathematics. Such a variant was anticipated, perhaps unsurprisingly, by … Kurt Gödel. Let's explore further.

2   "Seeing the truth" and "going outside the system"

The first to produce a full-fledged Gödelian argument against (strong) AI was the Oxford philosopher J.R. Lucas, in a 1961 paper called "Minds, Machines, and Gödel":

> Gödel's theorem seems to me to prove that Mechanism is false, that is, that minds cannot be explained as machines … Gödel's theorem must apply to cybernetical machines, because it is of the essence of being a machine, that it should be a concrete instantiation of a formal system. It follows that given any machine which is consistent and capable of doing simple arithmetic, there is a formula which it is incapable of producing as being true – i.e., the formula is unprovable-in-the-system – but which we can see to be true. It follows that no machine can be a complete or adequate model of the mind, that minds are essentially different from machines.[2]

---

[2]   Lucas (1961), pp. 43–4.

Hence Lucas drew some conclusions on the nature of human mind:

> However complicated a machine we construct, it will, if it is a machine, correspond to a formal system, which in turn will be liable to the Gödel procedure for finding a formula unprovable-in-that-system. This formula the machine will be unable to produce as being true, although a mind can see that it is true. And so the machine will still not be an adequate model of the mind. We are trying to produce a model of the mind which is essentially "dead" – but the mind, being in fact "alive," can always go one better than any formal, ossified, dead system can. Thanks to Gödel's theorem, the mind always has the last word.[3]

To understand the point, recall Gödel's claim that the undecidable sentence (say γ) of the formal system (say Typographical Number Theory) is "decided by metamathematical considerations." The informal argument, as you will remember, has it that γ "claims" (via arithmetization) to be not provable. So if it were provable, it would be false; therefore, if **TNT** is sound (it does not prove falsities), then γ must be unprovable in it. So γ is what it claims to be; therefore, γ is true. The argument cannot be formalized within **TNT**, for this would require truth-in-the-system to be definable within the system, which cannot be, because of Tarski's Theorem. So γ is decided in the metatheory of **TNT**, that is, in an environment supplying the machinery to define truth for (the language of) **TNT** by moving to the higher type, etc., etc.

This kind of jump into the metatheory has been associated with the phrase "going outside the system," which we have already met. This is an intriguing expression, and the Gödelian arguments in the philosophy of mind are likely to gain advantage from it. "Going outside the system" it consists in is what a system such as **TNT**, or a Turing machine, cannot do by definition. However, it seems that *we* can do what **TNT** cannot do: we seem to be able – to use another evocative term frequently employed in these contexts – to "transcend" the system, look at it from the outside, and thereby prove things the system cannot prove. Some say that getting out of the system – suspending the mechanical application of rules, and thus reasoning, so to speak, from an Ulterior Perspective – is what intelligence *consists in*.

[3] Ibid., p. 48.

In his famous 1989 book *The Emperor's New Mind*, and in its 1994 follow-up *Shadows of the Mind*, Roger Penrose – Rouse Ball Professor of Mathematics at Oxford, and one of the world's most creative mathematical physicists – has proposed in turn a Gödelian argument in order to establish the non-algorithmic nature of the human mind. The intents are much wider and more ambitious than Lucas'. Penrose is a committed Platonist in the philosophy of mathematics: he believes numbers to be abstract, objectively existing entities inhabiting their intelligible Platonic realm. And according to him Gödel's Theorem substantiates the Platonic idea of mathematical intuition: the idea, mentioned two chapters ago, that the mind can "see" certain mathematical truths with a kind of intellectual intuition (whereas, in particular, there are *no* algorithmic procedures to produce these truths, or to capture them within formal systems with their strict "finitary" constraints). According to Penrose, Gödel's argument allows us to surpass, by means of pure insight, any specific formalized mathematical system. Moreover, after showing that the human mind does not have an algorithmic nature, he believes it is possible to account for its nature by resorting to a "non-computational" extension of current quantum physics: consciousness should be explained by means of quantum interactions at the neural level.

The proof of the First Incompleteness Theorem given by Penrose in Chapter 4 of *The Emperor's New Mind* is a "semantic" one, establishing the unprovability of the Gödel sentence of the formal system at issue ("$P_k(k)$" in Penrose's notation) on the basis of the assumption that the system is sound. It follows that what $P_k(k)$ claims has to be true, and thus $P_k(k)$ is undecidable, since its negation is false, therefore unprovable, given the system's soundness. Penrose's conclusion is not too different from Lucas':

> In the course of the above argument, we have actually established that $P_k(k)$ is a true statement! Somehow we have managed to see that $P_k(k)$ is true despite the fact that it is not formally provable within the system. All this shows that the mental procedures whereby mathematicians arrive at their judgments of truth are not simply rooted in the procedures of some specific formal system. We see the validity of the Gödel proposition $P_k(k)$ though we cannot derive it from the axioms. The type of "seeing" that is involved in a reflection principle requires a mathematical insight that is not the result of the purely algorithmic operations that could be coded into some mathematical formal system.[4]

[4]   Penrose (1989), pp. 150 and 153–4.

3   The basic mistake

These arguments certainly warm people's hearts. Unfortunately, most commentators[5] agree that there is a mistake in them. The fault has to do with our seeing that the Gödel sentence is true. What (the first half of) the First Incompleteness Theorem states is that *if* **TNT** is consistent, *then* γ is unprovable in it. Now if γ is unprovable, then γ certainly is what it "claims" (via arithmetization) to be, so it is true. But to "see" that the Gödel sentence of the system is true, one has to "see" that **TNT** is consistent to begin with (or better still, sound). If **TNT** turned out to be inconsistent, then γ would be a false statement, since it claims to be unprovable in **TNT**, but **TNT** proves it: being an inconsistent system, it proves anything whatsoever, as usual, because of Scotus' Law. To conclude ("see") that γ is true, we need to prove the antecedent of the conditional claim "if Typographical Number Theory is consistent, then γ is not provable in it" which (the first half of) Gödel's First Theorem consists in: we need to recognize *that* **TNT** is consistent (or better still, sound).

We can make the point by resorting to the Second Theorem. As we know, in order to prove the Second Theorem one essentially establishes that the formula formally representing "If Typographical Number Theory is consistent, then γ is not provable in it" is a theorem of **TNT** itself. Thus our Typographical Number Theory, too, "knows," so to speak, that γ holds only on the assumption that Typographical Number Theory is consistent, in the following sense: the formal counterpart of that claim, that is, the formula $\mathbf{Cons_{TNT}} \rightarrow \gamma$, is a theorem of **TNT**. So far, **TNT** knows as much as we do about this story.

About 10 years ago, Lucas came back to his controversial argument by proposing the following amendment:

> There is a claim being seriously maintained by the mechanist that the mind can be represented by some machine. Before wasting time on the mechanist's claim, it is reasonable to ask him some questions about his machine to see whether his seriously maintained claim has serious backing. It is reasonable to ask him not only what the specification of the

[5]   See e.g. Putnam (1961), Chihara (1972), Lolli (2004). Putnam wasn't directly addressing Lucas' argument, but a partial anticipation of it found towards the end of Nagel and Newman (1958).

machine is, but whether it is consistent. Unless it is consistent, the claim will not get off the ground. If it is warranted to be consistent, then that gives the mind the premise it needs. The consistency of the machine is established not by the mathematical ability of the mind but on the word of the mechanist. The mechanist has claimed that his machine is consistent. If so, it cannot prove its Gödelian sentence, which the mind can none the less see to be true; if not, it is out of court anyhow.[6]

Unfortunately, the reassurance of the mechanist, i.e., of the supporter of AI, does not "give the mind the premise it needs" at all. If the mechanist believes that the machine is consistent, and we fully trust her, we will also believe that the machine is consistent. But to believe and to *know* – that is, to "see (and prove) as true" – are two different things. We believe the machine to be consistent; since we know (thanks to the first half of G1) that if the machine is consistent its Gödel sentence is true, then we can also believe that its Gödel sentence is true; but we don't *know* it yet.

But didn't I claim some pages ago that there exist mathematically perfectly respectable proofs of the consistency, and also of the soundness, of Typographical Number Theory? (And, a Platonist will add, isn't it manifest, when we look at the axioms of **TNT**, that they are true of the natural numbers, and therefore consistent?) Isn't it the case, then, that we do have evidence for the antecedent of the famous conditional – that we have evidence for the statement "**TNT** is consistent"? As soon as we know ("see") that **TNT** is consistent, we know ("see") the truth of γ. And we know that **TNT** cannot follow us down this path, for a consistency proof for **TNT** is not formalizable in **TNT**, because of the Second Theorem.

This is all well and good; however, the anti-AI Gödelian argument requires much more. It requires that, for any machine, and thus for any formal system satisfying the conditions of applicability of the Incompleteness Theorem, we can always "see" the truth of the relevant Gödel sentence. The anti-mechanist needs to phrase the argument in all generality, if she wants to maintain the irreducibility of the mind to any formal system or computing machine whatsoever (we didn't want the argument to establish only that we are smarter than a DVD player). Now, it seems that proving ("seeing," "recognizing") the consistency of

6   Lucas (1996), p. 117.

any such system is not simple at all; it is, on the contrary, something no person might be able to do. Let us see why.

4   In the haze of the transfinite

As you may recall, we can add γ to **TNT**, thereby obtaining a system **TNT**$_1$ = **TNT** + γ, in which γ is decidable by definition, since it is an axiom. You may also remember that **TNT**$_1$ is an incomplete system in its turn, with its own undecidable Gödel sentence γ$_1$. Adding γ$_1$ as an axiom, we have a system **TNT**$_2$ = **TNT**$_1$ + γ$_1$, ..., and so on. Now, says Lucas:

> The procedure whereby the Gödelian formula is constructed is a standard procedure – only so could we be sure that a Gödelian formula can be constructed for every formal system. But if it is a standard procedure, then a machine should be able to be programmed to carry it out too ... This would correspond to having a system with an additional rule of inference which allowed one to add, as a theorem, the Gödelian formula of the rest of the formal system, and then the Gödelian of this new, strengthened, formal system, and so on. It would be tantamount to adding to the original formal system an infinite sequence of axioms, each the Gödelian formula of the system hitherto obtained ... We might expect a mind, faced with a machine that possessed a Gödelizing operator, to take this into account, and out-Gödel the new machine, Gödelizing operator and all. This has, in fact, proved to be the case. Even if we adjoin to a formal system the infinite set of axioms consisting of the successive Gödelian formulae, the resulting system is still incomplete, and contains a formula which cannot be proved-in-the-system, although a rational being can, standing outside the system, see that it is true. We had expected this, for even if an infinite set of axioms were added, they would have to be specified by some finite rule or specification, and this further rule or specification could then be taken into account by a mind considering the enlarged formal system. In a sense, just because the mind has the last word, it can always pick a hole in any formal system presented to it as a model of its own workings. The mechanical model must be, in some sense, finite and definite: and then the mind can always go one better.[7]

[7]   Lucas (1961), pp. 48–9.

Let's have a closer look at this development of the argument. I have presented the operation of adding γ to **TNT**, γ_1 to **TNT**$_1$, γ_2 to **TNT**$_2$, and so on, precisely ... with an "and so on."This means that such an attempt to restore completeness, successively adding the various Gödel sentences, quickly begins to look predictable and mechanical. If it actually is predictable and mechanical, we appear to be able to recapitulate the whole process of adding infinitely many Gödel sentences to a formal system, or to a computing machine. We cannot specify the whole series extensively, of course, precisely because it is an infinite one. But we can sum it up in an axiom scheme (just as we summed up in the induction axiom of **TNT** an infinity of properties of natural numbers, albeit only a denumerable infinity).

Lucas' remark points at the fact that, all in all, the Gödel sentences are all "cast from one single mold,"[8] and can therefore be packed together in an axiom scheme. Call such a scheme γ_ω, and call **TNT**$_\omega$ the formal system obtained by adding the scheme to **TNT**. The Greek *omega* shows up because Cantor used it to designate the least transfinite ordinal number, that is, $\omega = \{0, 1, 2, 3, ...\}$ – where 0, 1, 2, 3, ... are taken as ordinals. Now, *also* **TNT**$_\omega$ has its Gödel sentence – say $\gamma_{\omega+1}$, which **TNT**$_\omega$ cannot decide despite its powerful axiom scheme. Hofstadter nicely puts the point thus:

> Any system, no matter how complex or tricky it is, can be Gödel-numbered, and then the notion of its proof-pairs can be defined – and this is the petard by which it is hoist. Once a system is well-defined, or "boxed," it becomes vulnerable.
>
> This principle is excellently illustrated by the Cantor diagonal trick, which finds an omitted real number for each well-defined list of reals between 0 and 1. It is the act of giving an explicit list, a "box" of reals, which causes the downfall ... Now in the case of formal systems, it is the act of giving an explicit recipe for what supposedly characterizes number-theoretical truth that causes the incompleteness. This is the crux of the problem with TNT + [γ_ω]. Once you insert all the [γ]'s in a well-defined way into TNT, there is seen to be some *other* [γ] –some unforeseen [γ] – which you didn't capture in your axiom schema.[9]

---

8   Hofstadter (1979), p. 462.
9   Ibid., p. 464.

So even after **TNT** has been strengthened by adding the γ_ω scheme, we can consider its Gödel sentence, $\gamma_{\omega+1}$, that **TNT**$_\omega$ cannot prove. This fact *seems* to strengthen the anti-mechanist's position. Things are not so simple, however. When one moves from the sequence $\gamma_1, \gamma_2, \ldots$ to the scheme γ_ω, an important "jump" has been made (in Cantorian terms: we have moved from the finite ordinals to the first transfinite ordinal, that is, ω). Of course, we can strengthen **TNT**$_\omega$ too by adding $\gamma_{\omega+1}$ as another axiom, and we obtain **TNT**$_{\omega+1}$. But **TNT**$_{\omega+1}$ has its Gödel sentence in its turn, say $\gamma_{\omega+2}$, and we rapidly find ourselves in another infinite chain of extensions; we can label them by means of the *successor ordinals* of ω, that is, $\omega + 1, \omega + 2, \ldots$, and so on. "And so on" means that we can sum up *this* new infinite series in a scheme, which we will label with the *limit ordinal* 2ω, i.e., $\omega + \omega$. Then come its successors, $2\omega + 1, 2\omega + 2, \ldots$. But we also have other limit ordinals, namely $3\omega, 4\omega, \ldots$, and also $\omega \times \omega = \omega^2$, then also $\omega^3, \omega^4, \ldots$, and then ω^ω.[10] Says Hofstadter:

> Then come some more extensions, some of whose names seem quite obvious, others of which are rather tricky. But eventually, we run out of names once again – at the point where the answer-schemas
>
> $\omega, \omega^\omega, \omega^{\omega^\omega}, \ldots$
>
> are all subsumed into one outrageously complex answer schema. The altogether new name "ε_0" is supplied for this one. And the reason a new name is needed is that some fundamentally new kind of step has been taken – a sort of irregularity has been encountered. Thus a new name must be applied *ad hoc*.[11]

We met ε_0 when I was talking of the Gentzen consistency proof for arithmetic. What makes it a novelty among novelties is that it is the first ordinal which cannot be obtained from ω by means of a finite number of additions, multiplications, and exponentiations. Now, there exists a theorem due to Alonzo Church and Stephen Kleene which maintains that there is no recursive notational system capable of assigning a name

---

[10]   For the technical details on this kind of progression, see Feferman (1962) – where in fact the addition of various consistency statements is considered.

[11]   Hofstadter (1979), p. 469.

to all constructive ordinals. And it is very doubtful that the human mind can actually go beyond such ordinals. Says Stewart Shapiro:

> In these terms, the Lucas–Penrose contest to write and assert Gödel sentences becomes a contest to enumerate recursive ordinals. One might think that all Lucas has to do is iterate the procedure of adding Gödel sentences (or the Feferman reflection principle) far enough. The problem, however, is with the crucial notion of "far enough". At some point, we are no longer sure we are on the right road. No machine can iterate the procedure through all and only the recursive ordinals. Can Lucas?[12]

According to Hofstadter, the right conclusion to draw from this situation is that "any human being simply will reach the limits of his own ability to Gödelize at some point," and "from there on out, formal systems of that complexity, though admittedly incomplete for the Gödel reason, will have as much power as that human being."[13]

One could similarly consider a procedure in which, instead of adding the Gödel sentences, we add the consistency statements for the formal systems – and we know that the Gödel sentence of a system to which the Incompleteness Theorem applies is typically provably (in the system) equivalent to its consistency statement. We add to a system **S** the formula formally representing "**S** is consistent." If we "recognize" **S** as consistent, we can also "recognize" as consistent S_ω, taken as an extension obtained by adding, by means of a scheme, infinitely many consistency statements. However, as we dig deeper and deeper into the transfinite, adding axioms and schemes and moving to more and more complex structures, our certainties concerning the consistency of the resulting systems start to fluctuate. As we have seen in the previous chapters, when we move to more powerful theories the respective consistency proofs typically become more complex; a proof of the consistency of **ZFC** may well lead us outside "ordinary mathematics." In general, if a system includes strong axioms of infinity, the human mind may not be capable of "seeing" and knowing its consistency at all. As soon as we don't know (anymore) that the system is consistent, we don't know (anymore) with certainty whether its Gödel sentence is true. All in all, it seems there is no ground

---

[12]   Shapiro (1998), p. 288.
[13]   Hofstadter (1979), p. 470.

for the claim that we can outdo any given formal system, and therefore any computing machine. But the Gödelian arguments required nothing less.

5   "Know thyself": Socrates and the inexhaustibility of mathematics

In *Shadows of the Mind*, the Gödelian argument of *The Emperor's New Mind* is expanded and reformulated in terms of Turing machines; but it encounters similar problems. In the third chapter of *Shadows* Penrose has proposed a seemingly different Gödelian argument, and commentators have started to talk of "Penrose's second argument," or "Penrose's new argument."The exposition is convoluted and a bit obscure, and critics (from Per Lindström to Stewart Shapiro)[14] disagree both on how to reconstruct it, and on where exactly the mistake is. Penrose himself, on the other hand, has declared in the online journal *Psyche* that such "second argument" is "not … the 'real' Gödelian reason for disbelieving that computationalism could ever provide an explanation for the mind."[15]

However, a different remark proposed by Penrose in *Shadows* is worth some attention.This concerns our incapacity to specify all of our mathematical abilities in terms of a formal system we can recognize as correct.The "Conclusion G" advanced on p. 76 of *Shadows* is:"Human mathematicians are not using a knowably sound algorithm in order to ascertain mathematical truth."This seems to express Penrose's usual thesis that human mind is not algorithmic. But instead of stressing the word "algorithm,"we could stress"knowably,"and say:"Human mind could not *know the soundness*" of such a formal system or algorithm. Now, this comes very close to something Gödel *did* claim – something that actually follows from the Incompleteness Theorem, and also reveals something important about us and our relation to mathematics. Let us have a look.

On December 26, 1951, Gödel gave one of the prestigious Gibbs Lectures at the American Mathematical Society. In previous years, the

---

[14]   See Lindström (2001), Shapiro (2003).
[15]   Penrose (1996), Section 4.1.

Lectures had hosted great mathematicians and scientists such as Hardy, von Neumann, Weyl, and Einstein, but Gödel was the first logician to be invited. The title of his lecture was: "Some Basic Theorems on the Foundations of Mathematics and Their Philosophical Implications." The "theorems" at issue were precisely the Incompleteness Theorems, and their "philosophical implications" concerned the nature of mathematics and the capacities of the human mind.

Now, Gödel claimed that what the Theorems do entail (specifically, the Second Theorem) is that mathematics is *inexhaustible*:

> It is *this* theorem [i.e., the Second Theorem] which makes the incomplet-ability of mathematics particularly evident. For, *it makes it impossible that someone should set up a certain well-defined system of axioms and rules and consistently make the following assertion about it: All of these axioms and rules I perceive (with mathematical certitude) to be correct, and moreover I believe that they contain all of mathemat-ics.* If someone makes such a statement he contradicts himself.[16]

Why? In the Gibbs Lecture Gödel made a distinction between "mathematics in the objective sense," that is, the totality of mathematical truths, and "mathematics in the subjective sense," that is, the set of mathematical truths provable or recognizable "with mathematical certitude." Now if someone recognizes ("knows with mathematical certitude") that a given formal system for mathematics – call it **S** again – is sound, she thereby knows that it is consistent. But if she knows that **S** is consistent, then she also knows that its consistency statement – say **Cons$_S$** – is true. On the other hand, **S** cannot prove its own consistency statement because of the Second Theorem. So the person knows an arithmetical truth, namely that represented by **Cons$_S$**, which the system doesn't prove. All in all, the system does *not* capture the totality of arithmetical truths, against what the person had claimed. But Gödel also added the following remark:

> However, one has to be careful in order to understand clearly the meaning of this state of affairs. Does it mean that no well-defined system of correct axioms can contain all of mathematics proper? It does, if by mathematics proper is understood the system of all true mathematical propositions; it does not, however, if one understands by it the system of all demonstra-ble mathematical propositions ... As to subjective mathematics, it is not

---

[16]   Gödel (1995), p. 309, italics in the original.

precluded that there should exist a finite rule producing all its evident axioms. However, if such a rule exists, we with our human understanding could certainly never know it to be such, that is, we could never know with mathematical certainty that all propositions it produces are correct.[17]

Understood in *this* sense, Penrose's Conclusion G, that "Human mathematicians are not using a knowably sound algorithm in order to ascertain mathematical truth," is correct. However, when so interpreted Conclusion G states not that our mind (or the human mathematician's mind) is not algorithmic, but that a formal system embodying all of our mathematical knowledge could not be *recognized* as correct by us. Conversely, *if* we recognize a formal system as a correct formalization of a part of our mathematics, we also know that it cannot be a formalization of the whole of mathematics, given that we can produce a sound extension of the system by adding its consistency statement, i.e., a mathematical truth the system cannot prove.

In the Gibbs Lecture, thus, Gödel acknowledged that G1 and G2 do not rule out the existence of an algorithmic procedure (a computing machine, an automated theorem prover) equivalent to the mind in the relevant sense – and he admitted that such a mechanism is compatible with mathematical Platonism. However, if such a procedure existed "we could never know with mathematical certainty that all the propositions it produce[d were] correct." Consequently, it may well be the case that "the human mind (in the realm of pure mathematics) [is] equivalent to a finite machine that … is unable to understand completely its own functioning": a machine too complex to analyze itself up to the point of establishing the correctness of its own procedures. Gödel inferred that what follows from the incompleteness results is, at most, a disjunctive conclusion:

> *Either mathematics is incompletable in this sense, that its evident axioms can never be comprised in a finite rule, that is to say, the human mind (even within the real of pure mathematics) infinitely surpasses the powers of any finite machine, or else there exist absolutely unsolvable Diophantine problems of the type specified* … It is this mathematically established fact which seems to me of great philosophical interest.[18]

---

[17] Ibid.
[18] Ibid, p. 310, italics in the original.

In other words, either the mind actually has a non-algorithmic and not fully "mechanizable" nature, or else there exist absolutely undecidable mathematical problems. But G1 and G2 don't allow us to go further and conclude that the true disjunct is the first one. According to Gödel, then, what follows from G1, and especially from G2, is that if our mind is a computing machine, it is one such that it "is unable to understand completely its own functioning." If we are indeed just Turing machines, then we cannot know exactly which Turing machines we are – and this is altogether a most suggestive conclusion. As Paul Benacerraf said in "God, the Devil, and Gödel": "If I am a Turing machine, then I am barred by my very nature from obeying Socrates' profound philosophical injunction: KNOW THYSELF."[19]

---

[19]   Benacerraf (1967), p. 30.

12

Gödel versus Wittgenstein and the Paraconsistent Interpretation

Some people believe Wittgenstein to be, quite simply, the greatest twentieth-century philosopher. However, philosophers argue about everything, and one of the things they argue about is the identity of the best among them. There is much more widespread agreement among logicians: from various polls taken at the end of the last century it turns out that Gödel definitely is the logician of the millennium (followed at a great distance by Alfred Tarski and Gottlob Frege; Saul Kripke is often mentioned in the subranking of the living logicians). Now since we are dealing with the philosophical interpretations of Gödel's Theorem, wouldn't it be nice if the greatest philosopher of our times had had something to say about the most important theorem proved by the greatest logician of our times?

In fact, this has actually happened. Wittgenstein's *Remarks on the Foundations of Mathematics* include some sparse comments on the Incompleteness Theorem. Wittgenstein has mainly and explicitly in his sights the First Theorem, but some of the things he says indirectly concern the Second too. On these comments, a lot of literature has been produced. To adequately address the issue one must consider the relations between Gödel's thought and Wittgenstein's; and doing this would require a whole book.[1] In the correlation between these two giants, in fact, not only the interpretation of a theorem is at issue, but also the very nature of mathematics, together with some philosophical questions of even greater scope.

---

[1]   And the best book I know on the subject … Is not a book, but the degree thesis of my friend and pupil Matteo Plebani (see Plebani (2007)).

1   When geniuses meet ...

Gödel and Wittgenstein were contemporary without ever meeting, but one can bet that, if they had met, their conversation would have been uneasy. Each wrote short notes on the other, rather cursorily, as if they wanted not to place too much importance on the other's work.[2] Gödel had been in close touch with the Vienna Circle, where Wittgenstein, for good and ill, had the charisma of a maharishi. But as Rebecca Goldstein has claimed, "neither could acknowledge the work of the other without renouncing what was most central in his own view. Each ... was a thorn deep in the other's metamathematics."[3]

What Goldstein refers to is the two geniuses' opposite conceptions of the very nature of mathematics. We know Gödel's Platonism well by now, and how it inspired him in his discovery (discovery, not construction) of the Theorem. On the other hand, a common trait between the so-called "early" and the so-called "later" Wittgenstein, and a point on which he never deflects, from the *Tractatus logico-philosophicus* to the final writings, is his profound aversion to metaphysical Platonism in general, and specifically in the philosophy of mathematics. Wittgenstein's anti-Platonism is atypical: on the one hand, it is irreducible to intuitionistic constructivism, to Hilbert's formalism (which was viewed by Wittgenstein as just the other side of the Platonic coin), and to the Vienna Circle's image of mathematics as "syntax of language," despite its having been so inspired by the *Tractatus* itself; on the other hand, it shares features with each of these positions.

It is likely that the most general common trait between Wittgenstein's different remarks on the Incompleteness Theorem consists in his attempt to separate what he calls the "proof," that is, the actual mathematical result, from what he calls the "prose," that is, the alleged philosophical consequences of the Theorem – especially those of a Platonic

---

[2]   Wittgenstein in the *Bemerkungen*:"It might justly be asked what importance Gödel's proof has for our work. For a piece of mathematics cannot solve a problem of the sort that trouble *us*" (Wittgenstein (1956), p. 177e). Gödel, in the famous unsent letter to Grandjean:"Wittg(enstein)'s views on the phil(osophy) of math had no inf(luence) on my work" (quoted in Goldstein (2005), p. 116).

[3]   Goldstein (2005), p. 90. "Metamathematics" here should be taken in a broad sense as "philosophy of mathematics": as we shall see in a moment, "metamathematics" in Hilbert's sense, for Wittgenstein, denotes nothing.

flavour. What Wittgenstein certainly rejected of an attitude he probably detected in Gödel, too, was the idea of establishing substantial philosophical theses, and even solving foundational problems, by means of a mathematical result.

Then again, reconstructing the whole Wittgensteinian philosophy of mathematics is far too difficult and would require too much space; so in this chapter I will deal only with what he had to claim on the Incompleteness Theorem – in particular, with one specific reading of what he claimed. The interpretation I have in mind can be referred to the recent development of *paraconsistent* logics and arithmetics. We've had a quick glance at these logics, characterized by the fact that various versions of Scotus' Law fail in them. The admission of contradictions within theories and formal systems having these as the underlying logics does not render such theories immediately trivial and useless: no logical chaos is expected, as would happen with classical logic. I believe that the models of such theories, besides recapturing *some* of the insights underlying Wittgenstein's position on Gödel's Theorem (not all of them, as we shall see), help to make it more plausible in the light of contemporary mathematical logic.

2   The implausible Wittgenstein

"Plausible," I say. One has to admit at the outset that Wittgenstein hasn't enjoyed a very good press among most of those who have been dealing with his remarks on the Incompleteness Theorem. Early commentators, such as Alan Anderson, Michael Dummett, and Paul Bernays, agreed in dismissing the whole thing as an unfortunate episode in the career of a great philosopher. It seems that Wittgenstein had in his sights only the informal account of the Theorem, presented by Gödel in the introduction of his paper, and was misled by it (not that he was the only one: because of the misunderstandings it originated, Helmer said that that exposition "without any claim to complete precision" is the only mistake in Gödel's paper). It is claimed that Wittgenstein erroneously considered essential to the Gödelian procedure the natural language interpretation of the Gödel sentence, as claiming "I am not provable," as well as the reference to the meaning of the formulas, and to their truth.

Commentators were particularly struck by the fact that Wittgenstein seems to have taken the Gödel formula as a *paradoxical* sentence, not too different from the usual Liar – and Gödel's proof itself, therefore, as the deduction of an inconsistency:

> 11. Let us suppose I prove the unprovability (in Russell's system) of *P*; then by this proof I have proved *P*. Now if this proof were one in Russell's system – I should in this case have proved at once that it belonged and did not belong to Russell's system. – That is what comes of making up such sentences. But there is a contradiction here! – Well, then there is a contradiction here. Does it do any harm here? [4]

In a letter to Karl Menger, Gödel himself explicitly ascribes to Wittgenstein the erroneous interpretation of his celebrated proof as the deduction of a paradox:

> It is indeed clear from the passage that you cite that Wittgenstein did *not* understand [my Theorem] (or that he pretended not to understand it). He interprets it as a kind of logical paradox, while in fact it is just the opposite, namely a mathematical theorem within an absolutely uncontroversial part of mathematics (finitary number theory or combinatorics). [5]

Zermelo, Perelman, and probably Russell himself made similar mistakes in the interpretation of the First Theorem, in the years following the publication of Gödel's results. It is usually maintained that the error rests on a confusion between a theory and its metatheory, or between syntax and semantics,[6] which makes it impossible to understand the difference between the *truth* predicate, inexpressible, as we know (by Tarski's theorem), within the theory to which the First Theorem applies, and the *provability* predicate, which, on the contrary, is (weakly) expressible.[7] Until a few years ago, the discussion on Wittgenstein's

---

[4]   Wittgenstein (1956), p. 51e.
[5]   Quoted in Goldstein (2005), p. 118.
[6]   See Perelman (1936), who claimed that Gödel had just discovered a new logical paradox; see also Dawson (1984) on Russell, and on Zermelo's letter to Gödel on this issue. In the correspondence between the two mathematicians, Dawson points out, Zermelo "failed utterly to appreciate Gödel's distinctions between syntax and semantics" (p. 80).
[7]   Anderson (1958), p. 486, explicitly charges Wittgenstein of such confusion.

remarks seemed to be concluded by the trustworthy verdict of Gödel himself, who, in a letter to Abraham Robinson, stated that Wittgenstein "advance[d] a completely trivial and uninteresting misinterpretation" of the First Theorem.[8]

However, in recent years some interpreters have argued that it is possible to extract interesting philosophical theses from the comments of the *Bemerkungen*. Juliet Floyd and Hilary Putnam[9] have claimed that Wittgenstein's intuitions anticipate some metamathematical acquisitions concerning the non-standard models of arithmetic. Wittgenstein's further remarks on Gödel, recently published in CD-ROM format within the Bergen project, according to Victor Rodych show that he didn't consider the self-referential natural language interpretation of the Gödelian sentence essential to the proof of the First Theorem; on the contrary, he "correctly understood the number-theoretic nature of Gödel's proposition."[10] And the debate is nowadays lively and rapidly evolving, with authoritative commentators taking a stance on Wittgenstein's real thoughts on the subject in the most important international reviews – from the *Journal of Philosophy* to *Dialectica* and *Erkenntnis*.[11]

I also believe that no significant philosophical idea is past its use-by date. And in this chapter I will try to show that it is possible to provide an interpretation of Wittgenstein's position on Gödel's results, quite plausible and respectable in the light of contemporary mathematical logic – that is to say, precisely from the point of view from which the comments of the *Bemerkungen* were most severely attacked.

My reading, however, will not follow the line of the latest commentators. In particular, I'm not following the path of non-standard models, suggested by Floyd and Putnam – although we shall deal with other models of arithmetic, which definitely deserve to be called "non-standard." I will read Wittgenstein's stance on Gödel's First Theorem as conforming to the single, simple argument to be set out below; then interesting facts will follow for the philosophical significance of the incompleteness results. In addition, the "single argument" will also allow me to vindicate and support two other ideas which harmed Wittgenstein's reputation among mathematicians and logicians: (1) his

---

[8]   Quoted in Dawson (1984), p. 89.
[9]   See Floyd and Putnam (2000).
[10] Rodych (2002), p. 380.
[11] See also Hintikka (1999); Rodych (1999), (2003); Floyd (2001).

plain rejection of Hilbert's very idea of a metamathematics; and (2) the view that we should not dramatize the possibility that a calculus turns out to be inconsistent (a dramatization which, according to some, puzzled Wittgenstein precisely because he began to pay attention to the role of consistency proofs within Hilbert's strategy).[12] Wittgenstein's ideas on contradictions and consistency proofs were dismissed as absurdities by the same commentators who found his remarks on the First Theorem outrageous.[13]

The "single argument," therefore, will organically capture several fundamental intuitions at the core of Wittgenstein's philosophy of mathematics – although, eventually, it will not capture them all. For instance, I will have to exploit some ideas on the formalization of deductive theories, and some notions of model theory, which constitute established acquisitions of the current practice in mathematical logic, but which Wittgenstein would probably have rejected. Furthermore, I will not trust Wittgenstein's own declarations, according to which his remarks should not have any strictly mathematical import. On the contrary, my interpretation will entail a strong revisionism with respect to classical logic and classical mathematics.

3   "There is no metamathematics"

At the core of the "single argument" is the idea that, in maintaining an interpretation of Gödel's proof that made of it a paradoxical derivation, Wittgenstein was just being coherent with his bold move of rejecting the standard distinction between theory and metatheory (and therefore between formalized arithmetic and metamathematics).

Logicians have learned precisely from Gödel's results (and from Tarski's, on the undefinability of truth) to be much more careful than they had been before in distinguishing between theory and metatheory and between syntax and semantics. We may therefore forgive Gödel's contemporaries for being a bit sloppy on this. Unlike Zermelo and Perelman, however, Wittgenstein *knowingly* refused several aspects of such distinctions. For instance, during his entire philosophical career

---

[12]   See Marconi (1984).
[13]   See, again, Anderson (1958) and Bernays (1959).

he never had second thoughts on his rejection of Hilbert's metamathematics. This is clearly expressed in the *Philosophical Remarks* (i.e., in thoughts that were put together before the publication of Gödel's 1931 paper) and, most explicitly, in a paragraph of the *Philosophical Grammar* whose title is precisely "There is no metamathematics":

> 4. I said earlier "calculus is not a mathematical concept"; in other words, the word "calculus" is not a chesspiece that belongs to mathematics.
>
> There is no need for it to occur in mathematics. – If it is used in a calculus nonetheless, that doesn't make the calculus into a metacalculus; in such a case the word is just a chessman like all the others.
>
> Logic isn't metamathematics either; that is, work within the logical calculus can't bring to light essential truths about mathematics. Cf. here the "decision problem" and similar topics in modern mathematical logic
>
> …
>
> (Hilbert sets up rules of a particular calculus as rules of metamathematics.)[14]

That is to say: Hilbert's metamathematics is, in fact, nothing but mathematics. It is not a metacalculus, because there are no metacalculi: it is *just one more* calculus.

Discussing Wittgenstein's motivations for discarding Hilbert's conception of metamathematics would take too much space here. Roughly, they are closely connected to a rejection of the Platonic and, as it were, "referential" idea that mathematical sentences describe an independently existing domain – the "realm of numbers." Wittgenstein takes mathematics as a family of calculi, each of which is, so to speak, self-contained: what provides a mathematical sentence with its meaning is not its referring to a self-subsistent numerical reality, but the place it occupies in the net of the theorems and rules of its own system.

Now, if one follows this path, the conclusion that Gödel's proof is actually a paradox similar to the Liar follows ineluctably. Contrary to what Bernays claimed, the discussion of Gödel's results in the *Bemerkungen* does *not* "suffer from the defect that Gödel's quite explicit premises of the consistency of the considered formal system is ignored."[15] Bernays'

---

[14]   Wittgenstein (1969), pp. 296–7.
[15]   Bernays (1959), p. 523.

charge just begs the question against Wittgenstein, for the consistency of the relevant system is precisely what is called into question by Wittgenstein's reasoning.[16] Let us see why.

4 Proof and prose

We already know that the "syntactic" proof of G1 can go together with the "semantic" account, according to which, since γ claims (via arithmetization) to be unprovable, and we have proved that it is, γ is what it claims to be, and therefore γ is true. Were the truth predicate expressible in the theory, it would originate the Liar paradox, whereas the provability predicate is (weakly) expressible without lurking contradictions. Because of this, as we have seen in the chapter on Platonism, the incompleteness results are taken by some (including, probably, Gödel) as establishing a fundamental irreducibility of truth to provability, and a cornerstone of mathematical realism. This is how the Gödelian results have become "one of the great moving forces behind the modern resurgence of Platonism."[17]

But precisely this development of the First Theorem is what Wittgenstein would label as the metaphysical prose, illegitimately attached to the real mathematical proof. In the chapter devoted to the Platonic reading of the Theorem, we have also seen that according to some authors (Floyd and Putnam, or Lolli) G1 establishes only the undecidability of γ within **TNT**, whereas the further "semantic" conclusion that γ is also *true* would be a "metaphysical claim."[18] On the other hand, we know that "semantic" versions of the First Theorem are indeed available, and they are mathematically and logically quite respectable (I have mentioned Smullyan's). Two other and quite different aspects of the "semantic prose" were unacceptable, though, from a Wittgensteinian perspective: (1) the idea that sentence γ, which is syntactically undecidable within **TNT**, can nevertheless – as is usually said

---

[16] For a particularly clear statement of this issue, see Rodych (2002), pp. 384–5.
[17] Shanker (1988), p. 171.
[18] Floyd and Putnam (2000), p. 632.

"with a wave of hands"[19] – be shown to be true (of course, under the hypothesis of the consistency of **TNT**) on the basis of a *meta*theoretic argument conducted "outside" the formal system; and (2) the consequent, aforementioned discrepancy between provability in any system capable of expressing elementary arithmetic, and arithmetical truth.

(1)   As for the first point: the "semantic prose" is to the effect that γ, undecidable within **TNT**, is nevertheless decided by means of meta-mathematical considerations. How is γ's truth established, then? By a metamathematical proof of γ, that is, by means of a *detour* through the metatheory. We know that this was clearly stated by Gödel in the opening section of his paper, when he declared that "the proposition that is undecidable *in the system PM* still was decided by metamathematical considerations."[20]

Now, it was probably this claim that initially perplexed Wittgenstein, for in the *Philosophical Remarks* he had already observed:

> What is a proof of provability? It's different from the proof of proposition.
>
> And is a proof of provability perhaps the proof that a proposition makes sense? But then, such a proof would have to rest on *entirely different* principles from those on which the proof of the proposition rests. There cannot be a hierarchy of proofs!
>
> On the other hand there can't in any fundamental sense be such a thing as meta-mathematics. Everything must be of one type (or, what comes to the same thing, not of a type) ...
>
> Thus, it isn't enough to say that *p* is provable, what we must say is: provable according to a particular system.
>
> Further, the proposition doesn't assert that *p* is provable in the system *S*, but in *its own* system, the system of p. That *p* belongs to the system *S* cannot be asserted, but must show itself.
>
> You can't say *p* belongs to the system *S*; you can't ask which system *p* belongs to; you can't search for the system of p. Understanding *p* means understanding its system. If *p* appears to go over from one system into another, then *p* has, in reality, changed its sense.[21]

---

[19]   Priest (1979), p. 222.
[20]   Gödel (1931), p. 19.
[21]   Wittgenstein (1964), p. 180.

We shouldn't let the fact that Wittgenstein here adopts the Tractarian "saying/showing" terminology misguide us. The essential point is that, within this framework, it is not possible that the very same sentence (say γ) turns out to be expressible, but undecidable, in a formal system (say **TNT**) and demonstrably true (under the aforementioned consistency hypothesis) in a different system (the metasystem). Wittgenstein rejected the idea that a mathematical sentence has, to use Russellian jargon, such an external relation to its own proof. If a sentence is internally related to its proof, i.e., if, as Wittgenstein maintained, the proof establishes the very meaning of the proved sentence, then it is not possible for *the same* sentence (that is, for a sentence with the same meaning) to be undecidable in a formal system, but decided in a different system (the metasystem).[22]

(2)   As for the second point: following this bold general doctrine, Wittgenstein had to reject both the idea that a formal system can be incomplete, and the Platonic consequence that no formal system proving only arithmetical truths can prove all arithmetical truths. If proofs establish the meaning of mathematical sentences, then there cannot be incomplete systems, just as there cannot be incomplete *meanings*:

> The edifice of rules must be *complete*, if we are to work with a concept at all - *we cannot make any discoveries in syntax*. - For, only the group of rules *defines* the sense of our signs, and any alteration (e.g. supplementation) of the rules means an alteration of the sense ... Mathematics cannot be incomplete; any more than a *sense* can be incomplete.[23]

One may object that Wittgenstein here is collapsing different levels again: he is confusing a theory with what the theory describes.

---

[22]   Shanker explains Wittgenstein's view thus: "(Hilbert's) Programme could only be successful by assuming at the outset that it is logically possible to step outside - and above - a system while staying within the broad confines of mathematics and achieve from this superior vantage point what had proved impossible from within the system. But if the referential conception of mathematical propositions which lies at the heart of this picture is removed we are left with a collection of calculi all of which stand on an equal footing. Here is no hierarchy of calculi, just different - autonomous - calculi" (Shanker (1988), p. 213).

[23]   Wittgenstein (1964), pp. 182 and 188.

According to the Platonic interpretation of the incompleteness result, it is not arithmetic, in the sense of the "realm of natural numbers," which is incomplete. If we are Platonists, as Gödel certainly was, we shall take the "realm of numbers" as perfectly complete, with its properties distributed in a maximal and consistent way among numbers. It is just that this realm cannot be fully captured by any formal theory: it is formalized arithmetic which is incomplete, not the arithmetic reality (say the standard model N), which the theory was supposed to describe.

However, Wittgenstein intentionally opposed precisely this referential picture of mathematics, according to which the meaning of mathematical sentences consists in their referring to, and describing, an independently existing reality – the picture of "arithmetic as the natural history (mineralogy) of numbers," of which "our whole thinking is penetrated."[24] According to him, the meaning of a mathematical sentence is determined by the rules that govern its use in the calculus, and in particular by its own demonstration (which is why a demonstrative incompleteness in the theory would become *eo ipso* an incompleteness in meaning):

> A psychological disadvantage of proofs that construct *propositions* is that they easily make us forget that the *sense* of the result is not to be read off from this by itself, but from the *proof* … I am trying to say something like this: even if the proved mathematical proposition seems to point to a reality outside itself, still it is only the expression of acceptance of a new measure (of reality).[25]

Consequently, the Platonic separation between provability and truth also has to go. The remarks on the First Incompleteness Theorem in the *Bemerkungen* are resolute on this point:

> 5. Are there true propositions in Russell's system, which cannot be proved in his system? – What is called a true proposition in Russell's system, then?
>
> 6. For what does a proposition's *"being true"* mean? *"p" is true = p.* (that is the answer).[26]

---

[24] Wittgenstein (1956), p. 116e.
[25] Ibid., pp. 76e–77e.
[26] Wittgenstein (1956), p. 50e.

Here Wittgenstein is probably identifying (mathematical) truth with assertability,[27] and assertability in informal contexts can correspond to provability in the context of a formal system: to be assertable with respect to a formal system is just to be provable in it, that is, to be a theorem of the system. Therefore, Wittgenstein concludes:

> If, then, we ask in this sense: "Under what circumstances is a proposition asserted in Russell's game" the answer is: at the end of one of his proofs [i.e., as a theorem], or as a "fundamental law" (Pp.) [i.e., as an axiom – and, of course, axioms are theorems]. There is no other way in this system of employing asserted propositions in Russell's symbolism.
>
> 7. "But may there not be true propositions which are written in this symbolism, but are not provable in Russell's system?" – "True propositions," hence propositions which are true in *another* system, i.e. can rightly be asserted in another game ... [A] proposition which cannot be proved in Russell's system is "true" or "false" in a different sense from a proposition of *Principia mathematica*.[28]

In the end, " 'True in Russell's system' means, as was said: proved in Russell's system; and 'false in Russell's system' means: the opposite has been proved in Russell's system."[29] By identifying truth and provability, and by rejecting the very idea of metamathematics, Wittgenstein was opposing some established results of contemporary logic – or, better, of contemporary classical mathematics and classical logic – whereas his position has often been connected, e.g., by Dummett, Kreisel, Kielkopf, and others, to a strong mathematical constructivism and to the so-called "strict finitism."[30] This speaks against Wittgenstein's own claim, according to which "it is my task, not to attack Russell's logic from within, but from without," and "my task is not to talk about (e.g.) Gödel's proof, but to pass it by."[31] However, it is possible to introduce a single argument that, by reinterpreting Gödel's results in the light of

---

[27] On this point, see Rodych (1999), pp. 178–9.

[28] Wittgenstein (1956), p. 50e.

[29] Ibid., p. 51e. That at the core of Wittgenstein's rejection of the Platonistic prose associated with Gödel's proof is his identification of truth with provability has been argued in detail by Rodych and Shanker in various essays (see Rodych (1999), (2003); Shanker (1988)).

[30] See Dummett (1959), pp. 504–5; Bernays (1959), p. 519. See also Kielkopf (1970).

[31] Wittgenstein (1956), p. 174e.

Wittgenstein's general standpoint, gives to the latter an unexpected plausibility precisely from the point of view of modern *non*-classical mathematical logic. Let us see how.

5 The single argument

The paraconsistent interpretation of Gödel's Theorem takes the stage now. The strategy I shall propose to vindicate Wittgenstein's position exploits an argument put forward by Richard Routley and Graham Priest in various influential essays.[32] It has not been developed with Wittgenstein in mind, but it is extremely interesting for our purposes, because it allows us to interpret Gödel's proof precisely as a paradoxical derivation. The core idea is to see what happens when one tries to apply the First Incompleteness Theorem to the theory that captures *our intuitive, or naïve, notion of proof*.

By "naïve notion of proof" Routley and Priest apparently mean that underlying ordinary mathematical activity: "proof, as understood by mathematicians (not logicians), is that process of deductive argumentation by which we establish certain mathematical claims to be true."[33] Since Hilbert, formal logicians have learned to treat proofs as purely syntactic objects: sequences of strings of symbols, manipulated via transformation rules, etc. However, *proving* something, for a working mathematician, amounts to establishing that some sentence is *true*.

Now, when we want to settle the question of whether some mathematical sentence is true or false, we try to deduce it, or its negation, from other mathematical sentences which are already known to be true. The process cannot go backwards *in infinitum*, though. We should therefore reach, eventually, mathematical sentences which are known to be true without having to be proved – e.g., because they are "self-evident." However, this is not important (nor is it important to establish which are the primal truths; concerning arithmetic, they may be, for instance, principles such as those of Peano, that is, claims according to which every number has a successor).

---

[32] Routley (1979); Priest (1979), (1984), (1987). The following exposition draws on Berto (2007b), Ch. 4.

[33] Priest (1987), p. 50.

It is clear that the naïve-intuitive theory that Routley and Priest link to the naïve-intuitive notion of proof is rather informal. However, "it is accepted by mathematicians that informal mathematics could be formalized if there were ever a point to doing so, and the belief seems quite legitimate."[34] Admittedly, this is a step the so-called later Wittgenstein, who disliked formalizations, may have questioned:

> The curse of the invasion of mathematics by mathematical logic is that now any proposition can be represented in a mathematical symbolism, and this makes us feel obliged to understand it. Although of course this method of writing is nothing but the translation of vague ordinary prose.[35]

However, we may reasonably assume that, when Wittgenstein made such claims, he was not questioning formalization itself, but the overwhelming importance attributed to it by philosophers and logicians looking for the "ideal language." On the contrary, we are now assuming precisely that formalization is nothing but the "translation of vague ordinary prose": one may regiment the fragment of English in which the naïve theory is expressed, and turn it into a formal language. Then, the primal truths may be written down in the (now) formalized language and taken, say, as axioms; and proofs may be expressed as formal arguments. After having been so translated, the naïve theory would certainly be sufficiently strong in the sense explained above, i.e., capable of representing all the (primitive) recursive functions.

But is the naïve notion of proof decidable, i.e., given Church's Thesis, recursive? This is much less straightforward, and it is likely that the crux of the argument lies here. By assuming that the proof relation of naïve arithmetic is recursive, the argument certainly challenges the standard perspective, which is taken as established precisely by Gödel's results. Moreover, the recognized standards for accepting a deductive argument as a proof within mathematics may vary across time, so the very naïve notion of proof may *change*. Gödel himself is considered as having suggested that *proof* may not be recursive because now and then we add new principles or rules to the old ones, after having ascertained that they entail various things accepted as true, and nothing known to be

---

[34]   Ibid., p. 51.
[35]   Wittgenstein (1956), p. 155e.

false. The way in which we perform such additions would not be itself rule-governed.

To this, Routley and Priest have two main replies. The first simply proposes eliminating the alleged diachronic variability by restricting our attention to the naïve notion of proof that conforms to the *current* standards.[36] The second, and more radical, is that Gödel's hypothesis may be refuted by the fact that the naïve notion of proof is *socially* learnt and taught. That the naïve notion of proof is decidable means that we can in principle effectively recognize a naïve proof when we see one. Now, Priest stresses, "it is part of the very notion of proof that a proof should be effectively recognizable as such"[37] – for the point of a naïve proof is that it is a way of settling the issue of whether a given mathematical claim is true or not. As Alonzo Church claimed:

> Consider the situation which arises if the notion of proof is non-effective. There is then no certain means by which, when a sequence of formulas has been put forward as a proof, the auditor may determine whether it is in fact a proof. Therefore he may fairly demand a proof, in any given case, that the sequence of formulas put forward is a proof; and until the supplementary proof is provided, he may refuse to be convinced that the alleged theorem is proved. This supplementary proof ought to be regarded, it seems, as part of the whole proof of the theorem.[38]

Moreover, by acknowledging that the naïve proof relation is decidable we can explain how we *learn* arithmetic – that is, via an effective procedure:

> We appear to obtain our grasp of arithmetic by learning a set of basic and effective procedures for counting, adding, etc.; in other words, by knowledge encoded in a decidable set of axioms. If this is right, then arithmetic truth would seem to be just what is determined by these procedures. It must therefore be axiomatic. If it is not, the situation is very puzzling. The only real alternative seems to be Platonism, together with the possession of some kind of sixth sense, "mathematical intuition."[39]

---

[36] See Priest (1987), p. 54.
[37] Priest (1987), p. 41.
[38] Church (1956), p. 53.
[39] Priest (1994), p. 343.

This point, too, meets some Wittgensteinian concerns on teaching and learning mathematical calculi as a public, social phenomenon. Perhaps the most amazing fact about mathematics as a discipline is the unanimity (generally speaking) of mathematicians on what counts as a proof. As Wittgenstein remarked, the whole "language game" of mathematical proofs would be rendered impossible by lack of consensus among mathematicians. If the notion of arithmetic proof were not effectively recognizable, then the process whereby mathematics is learnt, and the general agreement of working mathematicians on what counts as a mathematical proof, would turn out to be a mystery (of course, this is but a particular case of a famous, more general argument to the effect that *language* can only be learnt recursively, and so the grammar of a learnable language must be generated by a decidable set of rules).[40] On the contrary, as Routley claims, if the truths of mathematics are effective or effectively enumerable we can understand "how one generation of mathematicians learns what counts as true from the previous generation, namely they learn certain basic mathematical truths and how to prove others by making deductions."[41]

Of course, one can speak against the decidability of the naïve notion of proof on the basis of Gödel's results themselves. But one may argue that, in the context, this would beg the question against paraconsistentists – and against Wittgenstein too. *Both* Wittgenstein and the paraconsistentists, on the one side, *and* the followers of the standard view on the other, agree on the following thesis: the decidability of the notion of proof and its consistency are incompatible. But to infer from this that the naïve notion of proof is not decidable invokes the indispensability of consistency, which is exactly what Wittgenstein and the paraconsistent argument call into question.

Whatever we think of these replies, the effective decidability of the naïve notion of proof was Wittgenstein's assumption too. Wittgenstein believed that the naïve (i.e., the working mathematician's) notion of proof had to be decidable (i.e., recursive, given Church's Thesis), for lack of decidability meant to him simply lack of mathematical *meaning*. Wittgenstein believed that everything had to be decidable in mathematics. So whether the naïve proof relation is effectively decidable or not, the argument coheres with Wittgenstein's position on this point too.

---

[40]   On which see, famously, Davidson (1984), Ch. 1.
[41]   Routley (1979), p. 327.

Let **T** be, then, the formalization of our naïve, intuitive mathematical theory. Assuming that **T**, just like **TNT**, is sufficiently strong and its notion of proof is decidable, *if* **T** is consistent, then Gödel's First Theorem applies: so there is a sentence, say φ, which is not a theorem of **T**, but which can be established as true via *a naïve proof,* and therefore *is* a theorem of **T**. Of course, anything that is naïvely-intuitively provable is provable within the naïve-intuitive theory! So "assuming its consistency, it would, therefore, seem to be both complete and incomplete in the relevant sense."[42]

Now we have no way to avoid a contradiction: either we accept this one, i.e., ⊢$_T$ φ and ⊬$_T$ φ (which is quite close to Wittgenstein's remark, quoted a few pages ago: "let us suppose I prove the unprovability (in Russell's system) of *P*; then by this proof I have proved *P*. Now if this proof were one in Russell's system – I should in this case have proved at once that it belonged and did not belong to Russell's system"); or we have to admit that our naïve mathematical theory, with its naïve notion of proof, is *inconsistent.*

Specifically, Routley and Priest claim that, under a few quite plausible assumptions, the Gödel sentence φ for the (formalization of the) naïve theory can be proved within **T** itself, together with its negation – so one of the inconsistencies hosted by **T** is to the effect that ⊢$_T$ φ and ⊢$_T$ ¬φ. The philosophical point is that "This sentence is not provable" now has its "provable" understood as meaning "demonstrably true," and, as Wittgenstein conjectured, Gödel's proof becomes the derivation of a real paradox:[43]

> In fact, in this context the Gödel sentence becomes a recognisably paradoxical sentence. In informal terms, the paradox is this. Consider the sentence "This sentence is not provably true." Suppose the sentence is false. Then it is provably true, and hence true. By *reductio* it is true. Moreover, we have just proved this. Hence it is provably true. And since it is true, it is not provably true. Contradiction. This paradox is not the only one forthcoming in the theory. For, as the theory can prove its own

---

[42]   Priest (1984), p. 165.

[43]   Therefore, Anderson's comment on Wittgenstein, that "the conclusion to draw would not be that *P* at once 'belonged and did not belong' to Russell's system, but rather that Russell's system was inconsistent" (Anderson (1958), p. 488) is really of little importance: either horn of the dilemma makes us end up in a contradiction; and, as we shall see very soon, *both* contradictions (i.e., a system proving both its Gödel sentence and its negation, and a system both proving and not proving something) are expected in a thoroughly paraconsistent framework.

soundness, it must be capable of giving its own semantics. In particular, [every instance of] the T-scheme for the language of the theory is provable in the theory. Hence ... the semantic paradoxes will all be provable in the theory. Gödel's "paradox" is just a special case of this.[44]

I have claimed that the "semantic prose" on the First Theorem attacked by Wittgenstein has it that the truth of the Gödel sentence is established in the metatheory (under the assumption that the theory is consistent): it can be proved in a metatheoretical environment in which we can deal with the semantics of the object theory, i.e., with the truth predicate for (the language of) the object theory. However, **T**, formalizing as it does our naïve notion of proof, should absorb the metatheory within the theory. After all, as Wittgenstein might have added, mathematicians use ordinary English, and ordinary English may well be (and, according to some philosophers of language, actually is) semantically closed. As Routley stressed, "everyday arithmetic as presented within a natural language like English appears, unlike say first-order Peano arithmetic, appropriately closed." And "is provable in arithmetic" and "is arithmetically true" are "English, and in a good sense arithmetical, predicates."[45]

So **T** is semantically closed in the Tarskian sense, and inconsistent. The reasoning behind the proof of the truth of the Gödel sentence is now performed within the formal system itself – which is what we should expect in a Wittgensteinian framework that collapses, in the aforementioned sense, the distinction between theory and metatheory. There is no metasystem in which one establishes that (if the object system is consistent, then) the Gödel sentence is true: there are no metasystems. Consequently, one cannot "get out of" a system and solve, in its metasystem, problems that were meaningfully expressible but undecidable within the system.

6   But how can arithmetic be inconsistent?

After this draft of the "single argument," one may reasonably ask for more information on what kind of theory **T** is. One will most likely

[44]   Priest (1987), pp. 46–7; see also Priest (1984), p. 172.
[45]   Routley (1979), p. 326.

wonder how Wittgenstein's position can be made more palatable from a logical point of view by referring it to an inconsistent theory. It is easy to see how audacious the argument itself is: by turning Gödel's proof into a paradox, it places inconsistencies at the very core of (the theory which, supposedly, captures) our mathematical practice.

Things seem to get worse if we follow a development of the argument due to Shapiro. Shapiro has proposed to refine the Priest–Routley notion of "provability in the naïve theory," which Chihara and others found less than precise,[46] by building a semantically closed formal arithmetic (say **TNT+**) including its own truth predicate. The theory proves its Gödel sentence (say $\gamma+$) and the truth predicate, as expected, is used in the proof. However, $\gamma+$ is itself purely arithmetic and, given the soundness of **TNT+**, embarrassing consequences follow: some number n both is and is not the code of a proof of $\gamma+$ and, more generally, it turns out not only that the theory proves both something and its negation, but that something could be both provable and not provable[47] – which is hard to swallow if anything is. Is this not the purest *reductio ad absurdum* of Wittgenstein's ideas on Gödel's theorem, and of his tentative rejection of metamathematics?

This is not so straightforward, though, unless one believes, like Wittgenstein's early commentators, that "contradictions do make formal systems uninteresting."[48] Here comes into play the aspect of Wittgenstein's philosophy of mathematics mentioned above, which my interpretation can recapture, namely, his attitude towards contradictions. That he did not consider the surfacing of contradictions within formal systems as a terrible crisis is well known and is testified, for instance, by his long discussions with Turing on this point, as reported in the *Lectures on the Foundations of Mathematics*.

It is true that Wittgenstein did not comment directly on Gödel's Second Incompleteness Theorem. However, he often commented on the role and importance of consistency proofs; and his position was clear-cut: he considered this kind of proof as a symptom of "the superstitious fear and awe of mathematicians in face of contradiction."[49]

[46] See Chihara (1984).
[47] See Shapiro (2002), p. 817.
[48] Anderson (1958), p. 484.
[49] Wittgenstein (1956), p. 53e.

After interpreting Gödel's proof as a paradox closely related to the Liar, Wittgenstein asks, rhetorically: "but there is a contradiction here! – Well, then there is a contradiction here. Does it do any harm here?" In Victor Rodych's words: "'perhaps', Wittgenstein might say, 'all calculi that admit such sentence-constructions are syntactically inconsistent'";[50] but he believed that a calculus within which one can derive a contradiction does not thereby become useless:

> Can we say: "Contradiction is harmless if it can be sealed off"? But what prevents us from sealing it off? ...
> Let us imagine having been taught Frege's calculus, contradiction and all. But the contradiction is not presented as a disease. It is, rather, an accepted part of the calculus, and we calculate with it ...
> For might we not possibly have wanted to produce a contradiction? Have said – with pride in a mathematical discovery: "Look, this is how we produce a contradiction"? ...
> My aim is to alter the *attitude* to contradiction and to consistency proofs. (*Not* to show that this proof shows something unimportant. How *could* that be so?)[51]

Because of these insights, Wittgenstein has been considered a forerunner of paraconsistent logics. He anticipated the intuition that, by rejecting Scotus' Law (that is, as we know, the classically and intuitionistically valid logical law according to which a contradiction entails everything), we may admit that an inconsistent calculus does not thereby become trivial and uninteresting:[52]

> And suppose the contradiction [i.e., Russell's paradox] had been discovered but we were not excited about it, and had settled e.g. that no conclusions were to be drawn from it.[53]

In the dialogues with Turing:

> *Turing*: If you say that contradictions will not really lead one into trouble, you seem to mean that one will take up towards contradictions the attitude which I described.

---

50   Rodych (1999), p. 190.
51   Wittgenstein (1956), pp. 104e–106e.
52   On this point, see Marconi (1984).
53   Wittgenstein (1956), p. 170e.

Wittgenstein:You might get *p*. ~ *p* by means of Frege's system. If you can draw any conclusion you like from it, then that, as far as I can see, is all the trouble you can get into. And I would say, "Well then, just don't draw any conclusions from a contradiction."[54]

Now, if we adopt a paraconsistent logic then the theory **T** mentioned above, which according to Routley and Priest captures our naïve-intuitive notion of proof, is not just an argumentative trick anymore. It is possible to provide a respectable logical framework for Wittgenstein's idea according to which Gödel's proof is paradoxical, and nevertheless the derivation of paradoxes of the Gödelian kind does not render the relevant system(s) useless. Inconsistent arithmetics, i.e., non-classical arithmetics based on some paraconsistent logic or other, are nowadays a reality, both from the proof-theoretic and from the model-theoretic point of view. What is more important, the theoretical features of such theories precisely match some of the aforementioned Wittgensteinian intuitions. Let us see a few examples of such correspondences.

First, paraconsistent arithmetics do not fulfill precisely the consistency requisite, which is a precondition for the applicability of many limitative theorems. This suggests that such theories could, in a sense, emancipate themselves from Gödel's Theorems, and maybe from other limitative results afflicting their consistent cousins based upon a more traditional (classical, or intuitionistic) logic. Of course, consistency proofs are out of question, since we are dealing with inconsistent theories. What the theory may hopefully prove, though, is its own non-triviality (that is, the fact that it does not prove everything), which in these contexts is more often called *absolute* consistency. Paraconsistent authors have begun to show that this is the case since the 1970s, by building inconsistent but non-trivial formal systems whose non-triviality proof can be represented within the very systems and that, in this sense, circumvent to some extent Gödel's Second Theorem. Their inconsistency allows them to escape also from Gödel's *First* Theorem, and from Church's undecidability result: they are, that is, demonstrably complete and decidable.[55] They therefore fulfill precisely Wittgenstein's

---

[54] Wittgenstein (1976), p. 220.
[55] I shall not enter into the technical details of these remarkable results here. For a quick presentation, see Bremer (2005), Ch. 13, and Berto (2007b), Ch. 11. For non-triviality proofs for paraconsistent set theories, see Brady (1989), Brady and Routley (1989).

request, according to which any authentically mathematical calculus should be complete: there should not be mathematical problems that can be meaningfully formulated within the system, but which the rules of the system cannot decide. Hence, the decidability of paraconsistent arithmetics harmonizes with an opinion Wittgenstein maintained throughout his philosophical career.

Furthermore, the perspective of inconsistent arithmetics is (typically, though not necessarily) involved in a form of strict *finitism*. Some of the models of inconsistent arithmetics have a smaller cardinality than that of the standard model $\mathbb{N}$. Such models can be obtained by applying to $\mathbb{N}$ an appropriate filter that reduces its cardinality. Meyer and Mortensen have initially developed the technique, and some of their main results are summarized in a paper that appeared in the *Journal of Symbolic Logic*, in which different finite models are considered.[56] In general, the filter works as follows. Let D be the domain of a given model $\mathbb{M}$, and $\approx$ an equivalence relation defined on D, which is also a congruence with respect to the denotations of the function symbols of the language. Given the objects $o_1, \ldots, o_n$ belonging to D, $|o_1|, \ldots, |o_n|$ are the corresponding equivalence classes under $\approx$. Now, let $\mathbb{M}^{\approx}$ be the new model, called the *collapsed model*, whose domain is $D^{\approx} = \{|o| \mid o \in D\}$. $\mathbb{M}^{\approx}$ has the role of providing substitutes for the initial objects, and particularly of identifying the members of D in each equivalence class, thereby producing a composite object that "inherits" the properties of its components: the predicates that were true of the initial objects now apply to the substitute.

Now, by induction over the complexity of formulas it is possible to prove the following lemma, called the Collapsing Lemma:

(CL)   Given any formula α which has the truth value v in M, α has the truth value v also in $M^{\approx}$.[57]

Therefore, if the original model satisfied some set of formulas, the collapsed model also satisfies it: when the initial model $\mathbb{M}$ is collapsed into $\mathbb{M}^{\approx}$, no sentence loses a truth value: it can only gain them. Of course, when we begin with the model of a standard theory, the only values around are *true* and *false*. But in the collapsed model it may be the case

---

[56]   See Meyer and Mortensen (1984).
[57]   See Priest (1994), pp. 346-7; the result was anticipated in Dunn (1979).

that a formula, which was initially true only, or false only, becomes both true and false, that is, paradoxical (a truth value admitted in the semantics of some paraconsistent logics, such as **LP**, the *logic of paradox*: **LP** is a three-valued logic whose values are precisely *true only*, *false only*, and *both true and false* or *paradoxical*).[58] This happens when the collapsing filter produces an inconsistent object: for instance, it may identify in an equivalence class two initial objects, one of which had, whereas the other didn't have, the very same property. The procedure works even if among the relevant sentences we have formulas that seem to put constraints on cardinality, such as $\exists x \exists y (x \neq y)$, precisely because *they* can become paradoxical.

The underlying intuition would be that there is a finite (albeit hardly imaginable and unknown to us) number of things in the world. Although we cannot specify the number, we know that it must be "a number larger than the number of combinations of fundamental particles in the cosmos, larger than any number that could be sensibly specified in a lifetime"[59] (which should explain why our intuitions on it are rather unconfident); and this largest number is an *inconsistent* number.[60]

Now, suppose that n is our largest inconsistent number. Let **N** be the theory of $\mathbb{N}$, that is, the set of arithmetical sentences true in the standard model $\mathbb{N}$ (corresponding to what we have called the True Arithmetic in previous chapters); and let $\mathbf{M}_n$ be the set of sentences true in the paraconsistent model with the inconsistent number n. We may take as the underlying logic of $\mathbf{M}_n$ some mainstream paraconsistent logic, such as **LP**. Now, according to Priest[61] such a theory as $\mathbf{M}_n$ has the following enjoyable properties: it is, of course, inconsistent (including, among other things, both its own Gödel sentence and its negation), but provably non-trivial – and its non-triviality proof can be formalized within it. It fully contains **N**, that is, it includes all the sentences true in the standard model.[62] Finally, $\mathbf{M}_n$ includes its own truth predicate. Therefore, the

---

[58] See e.g. Priest (1979), (1987).

[59] Priest (1994), p. 338.

[60] Such a strict finitism is not unavoidably tied to inconsistency, nonetheless: van Bendegem (1994), (1999) has exploited the properties of paraconsistent arithmetical models to argue for a largest number which is not an inconsistent one.

[61] See Priest (1994), p. 337.

[62] In the case of $\mathbf{M}_n$, the trick consists in choosing for $\mathbb{N}$ a filter that (a) given a number $x < n$, puts x and nothing else in the corresponding equivalence class, so that $|x|$ inherits

inconsistent arithmetic avoids Gödel's First IncompletenessTheorem; it also avoids the SecondTheorem, in the sense that its non-triviality can be established within the theory; and Tarski's Theorem too – including its own predicate is not a problem for an inconsistent theory.[63]

Priest has declared that the Collapsing Lemma (CL) is "the ultimate downwards Löwenheim–Skolem Theorem,"[64] which is easy to understand. The downward half of the Löwenheim–Skolem Theorem, as you might recall, claims that any first-order theory, which has a model with an infinite domain, has a model with a denumerably infinite domain too. The filter and the Collapsing Lemma allow us to "shrink" even more, since one can reduce a model with a denumerably infinite domain into one of any smaller size. We can have a collapsed model $\mathbb{M}^=$ whose domain $D^=$ has cardinality k (smaller than that of the initial model), by choosing an appropriate equivalence relation that produces precisely k equivalence classes. Bremer has therefore suggested the following Paraconsistent Löwenheim–Skolem Theorem: "Any mathematical theory presented in first order logic has a *finite* paraconsistent model."[65]

Now, this strong finitism also meets a persistent tendency in Wittgenstein's philosophy of mathematics. Wittgenstein always had a negative attitude towards Cantor's paradise and the non-denumerable infinities which, in Cantor's Platonic view, were to be discovered by the diagonal argument. Of course, strict finitism and the insistence on the decidability of any meaningful mathematical question go hand in hand. As Rodych has remarked, the intermediate Wittgenstein's view is dominated by "his finitism and his ... view of mathematical meaningfulness as algorithmic decidability," according to which "[only] finite logical

---

all and only the properties of x; and (b) puts every number $y \geq n$ in a single equivalence class. Consequently, all the true/false equations involving any number smaller than n in the standard model are now true only or false only of the substitute. Because of this, the initial segment in the succession (which is sometimes called the *tail*) behaves as usual. Roughly, "up to n" things work as in ordinary arithmetic. Nevertheless, anything that could be truly/falsely claimed of anything bigger than n is now true/false of the inconsistent number. Many things concerning it are therefore paradoxical now (both true and false), and n is an inconsistent object. Specifically, in the model $\mathbf{n} = \mathbf{n} + 1$ is true even though it is also false.

[63]   For a detailed account of these facts, see Priest (1994), pp. 337–8; see also Priest (1987), pp. 234–7.

[64]   Priest (1994), p. 339.

[65]   Bremer (2005), p. 155.

sums and products (containing only decidable arithmetic predicates) *are* meaningful because they are *algorithmically decidable.*" But this tendency remains also in the later phase: "As in the middle period, the later Wittgenstein seems to maintain that an expression is a meaningful proposition only *within* a given calculus, and *iff* we knowingly have in hand an applicable and effective DP [decision procedure] by means of which we can decide it." [66]

7 The costs and benefits of making Wittgenstein plausible

The cost of accepting paraconsistent arithmetics is clear: we have to revise some well-established results of classical mathematical logic. As I have claimed before, by subscribing to such a way of making sense of Wittgenstein's remarks on Gödel's Theorem we will not be allowed to claim – as many commentators did – that such a philosophy does not require any logical-mathematical revisionism, being directed only against the foundational demands of philosophers.

On the other hand, I believe that Wittgenstein might have found the situation produced by paraconsistent arithmetics quite plausible. Surprising and (in a broad sense) paradoxical innovations in the history of mathematics – this "motley of techniques of proof" [67] – led to new kinds of numbers: from Hyppasus' irrational numbers refuting Pythagorism, to infinitesimals, Cantor's transfinite numbers, imaginary numbers, surreal numbers, and all that. The early reception of such new entities among mathematicians has often been controversial, from the Pythagoreans condemning and expelling Hyppasus, to Kronecker making Cantor's life impossible. A process of rethinking mathematics in order to come to grips with the new domain has usually followed. And, as we have seen, such an audacious rethinking in a paraconsistent framework may nowadays vindicate some of Wittgenstein's "outrageous claims" on Gödel's Theorem, too swiftly dismissed by commentators who dogmatically took the logic of Russell and Frege as the One True Logic.

[66] See Rodych (1999), pp. 174–6.
[67] Wittgenstein (1956), p. 84e.

Epilogue

The internal paradoxes in Gödel's personality were at least partially provoked by the world's paradoxical responses to his famous work. His incompleteness theorems were simultaneously celebrated and ignored. Their technical content transformed the fields of logic and mathematics; the method of proof he used, the concepts he defined in the course of the proof, led to entirely new areas of research, such as recursion theory and model theory ... Yet, the metamathematical import of the theorems, which to Gödel was their most important aspect, was disregarded. (Rebecca Goldstein, *Incompleteness*)

During the oral examination of Simon Kochen's doctorate, Professor Stephen Kleene, who was examining him, demanded: "Name me five results by Gödel." If I had been Kochen, I would have mentioned: the completeness of first-order classical logic; the proof of the consistency of the Continuum Hypothesis with the axioms of set theory; the hierarchy of constructible sets; the *Dialectica* interpretation; and the Incompleteness Theorem. In the book you are about to conclude, we have been dealing only with the last; but each of these results is a breakthrough in contemporary logic and mathematics and has opened still developing fields of research.

Gödel's work is so important that he is also celebrated for what he did not achieve but could have – and according to some actually did, but only in the solitude of his introspective mind. For some time he was suspected to have proved the independency of the Continuum Hypothesis before Paul Cohen, in the forties, and to have kept the proof to himself. He always denied this.[1] As Feferman has remarked, we find Gödel at the starting point of the two fundamental notions of contemporary logic: *truth* and *computability*.[2] He certainly anticipated Tarski's Theorem on the undefinability of truth, that is, the single most important result achieved by Tarski. But in fact Gödel never dealt officially and directly with the issue, probably because of the suspicions surrounding the concept of truth among his contemporaries. He certainly was not persuaded by Church's Thesis and contributed less than he could have done to the development of recursion theory (of which he remains, nevertheless, a co-founder), and thus to contemporary ideas on computability, computers, and artificial intelligence – maybe because he did not like the idea that the human mind might be just a Turing machine, although he was cautious enough to be aware of the fact that such an idea could not be refuted by his mathematical results.

Of all these amazing (completed or incomplete) endeavors, the incompleteness of arithmetics is the most remarkable. As has been hinted at several pages ago, some have claimed that the aforementioned incompletenesses in Gödel's work have their roots precisely in his philosophical convictions: his faith in the mind's capacity to "see" transfinite truths, his deep Platonism.

I believe the opposite is true. I think philosophy was the deep propulsive force of Gödel's thought. Even if the Incompleteness Theorem doesn't provide a foundation for mathematical Platonism, the latter certainly was the heuristic guide for establishing the former. Behind the diagonal procedure for obtaining self-referential formulas within formalized arithmetic we find the inspiration of the ancient Greek philosophers who lost their health thinking about the Liar paradox. And behind the idea of the Gödel numbering we find Leibniz's philosophy (an author Gödel loved and strongly identified with during a period of his life), with the attempted arithmetization of language in the *De arte combinatoria*.[3]

---

[1] See Dawson (1997), pp. 224–5.

[2] See Feferman (1983).

[3] Van Atten and Kennedy (2003) is a nice work on the philosophical development of the later Gödel, and especially on his transition from interest in Leibniz to involvement with Husserl.

Gödel believed in an "interesting axiom" – not a mathematical axiom, but a philosophical one: *Die Welt ist vernünftig*, "The world is intelligible."[4] The axiom is quite similar to the principle, dating back to Leibniz, which claims: *Nihil est sine ratione*. Those who are familiar with the Leibnizian metaphysics of monads – these simple substances that never interact causally with each other, but develop their intrinsic virtuosities, and which are coordinated by an incredible pre-established harmony – know how implausible it sounds: Russell called it a "fairy tale." Gödel, however, liked fairy tales: he loved Walt Disney movies, especially *Bambi* and *Snow White*, which he saw at least three times.[5] Three features shared by most tales are their simplicity, creativity, and capacity to carry some message. And the proof of the Incompleteness Theorem, besides resembling an "intellectual symphony," also reminds us of a tale. It is based on the two simple ideas we have become familiar with, namely: the Gödel numbering, which allows arithmetic to talk of quite unexpected things; and the diagonalization procedure, which allows an arithmetical formula to talk about itself. The combination of these two insights is an act of creativity you would never expect from a mathematical logician.

As for the message delivered by Gödel's incompleteness … well, after reading this book, you might agree on the claim that this is the tricky part of the story. Part of the decoding required to decipher the message may consist in this: since the world is intelligible, you can find meaningfulness even where you least expect it, for instance in a package of strings of symbols that turn out to be capable of introspection – and capable, furthermore, of doing what many human beings also do, namely get into trouble precisely *because of* their intelligence and self-awareness.

---

[4] See Goldstein (2005), p. 21.
[5] See Dawson (1997), p. 181.

References

Anderson A.R. (1958) "Mathematics and the 'Language Game'," in Benacerraf and Putnam (1964) pp. 481-90.

Anderson A.R. (ed.) (1964) *Minds and Machines*, Prentice Hall, Englewood Cliffs, NJ.

Aristotle, *The Basic Works*, ed. R. McKeon, Random House, New York.

Atten M. van, Kennedy E. (2003) "On the Philosophical Development of Kurt Gödel," *Bulletin of Symbolic Logic*, 9, pp. 425-76.

Beall J.C., Fraassen B. van (2003) *Possibilities and Paradox: An Introduction to Modal and Many-Valued Logic*, Oxford UP, Oxford.

Bellotti L., Moriconi E., Tesconi L. (2001) *Computabilità: Lambda-definibilità, ricorsività, indecidibilità*, Carocci, Rome.

Benacerraf P. (1967) "God, the Devil, and Gödel," *The Monist*, 51, pp. 9-32.

Benacerraf P., Putnam H. (eds) (1964) *Philosophy of Mathematics: Selected Readings*, Prentice Hall, Englewood Cliffs, NJ.

Bendegem J.-P. van (1994) "Strict Finitism as a Viable Alternative in the Foundations of Mathematics," *Logique et Analyse*, 137, pp. 23-40.

Bendegem J.-P. van (1999) "Why the Largest Number Imaginable Is Still a Finite Number," *Logique et Analyse*, 165-6, pp. 107-26.

Bernays P. (1959) "Comments on Ludwig Wittgenstein's *Remarks on the Foundations of Mathematics*," in Benacerraf and Putnam (1964) pp. 510-36.

Bernays P., Fraenkel A. (1958) *Axiomatic Set Theory*, North-Holland, Amsterdam.

Berto F. (2005) *Che cos'è la dialettica hegeliana?*, Poligrafo, Padua.

Berto F. (2006a) *Teorie dell'assurdo. I rivali del Principio di Non-Contraddizione*, Carocci, Rome.

Berto F. (2006b) "Meaning, Metaphysics and Contradiction," *American Philosophical Quarterly*, 43, pp. 283-97.

Berto F. (2007a) *Logica da zero a Gödel*, Laterza, Rome.

Berto F. (2007b) *How to Sell a Contradiction: The Logic and Metaphysics of Inconsistency*, College Publications, London.

Berto F. (2007c) "Is Dialetheism an Idealism? The Russellian Fallacy and the Dialetheist's Dilemma," *Dialectica*, 61, pp. 235-63.

Berto F. (2008) "'Αδύνατον and Material Exclusion," *Australasian Journal of Philosophy*, 86, pp. 165-90.

Boolos G. (1990) "On Seeing the Truth of the Gödel Sentence," *Behavioral and Brain Sciences*, 13, pp. 655-6.

Boolos G., Burgess J.P., Jeffrey R.C. (2002) *Computability and Logic*, 4th edn, Cambridge UP, Cambridge.

Brady R.T. (1989) "The Non-Triviality of Dialectical Set Theory," in Priest, Routley, and Norman (1989) pp. 437-71.

Brady R.T., Routley R. (1989) "The Non-Triviality of Extensional Dialectical Set Theory," in Priest, Routley, and Norman (1989) pp. 415-36.

Bremer M. (2005) *An Introduction to Paraconsistent Logics*, Lang, Frankfurt a.M.

Brouwer L.E.J. (1975) *Collected Works*, ed. A. Heyting, vol. I, North-Holland, Amsterdam.

Burali-Forti C. (1897) "Una questione sui numeri transfiniti," *Rendiconti del circolo matematico di Palermo*, 11, pp. 154-64.

Cantor G. (1895) "Beiträge zur Begründung der transfiniten Mengenlehre," *Mathematische Annalen*, 46, pp. 481-512, tr. in *Contributions to the Founding of the Theory of Transfinite Numbers*, Dover, New York, 1955.

Cantor G. (1899) "Cantor an Dedekind," in Cantor (1932) pp. 443-7.

Cantor G. (1932) *Gesammelte Abhandlungen mathematischen und philosophischen Inhalts*, ed. E. Zermelo, Springer, Berlin.

Carnap R. (1937) *The Logical Syntax of Language*, Routledge & Kegan Paul, London.

Chihara C. (1972) "On Alleged Refutations of Mechanism using Gödel's Incompleteness Theorems," *Journal of Philosophy*, 69, pp. 507-26.

Chihara C. (1984) "Priest, the Liar, and Gödel," *Journal of Philosophical Logic*, 13, pp. 117-24.

Church A. (1956) *Introduction to Mathematical Logic*, Princeton UP, Princeton, NJ.

Davidson D. (1984) *Inquiries into Truth and Interpretation*, Oxford UP, Oxford.

Davis M. (ed.) (1965) *The Undecidable*, Raven, New York.

Davis M. (1982) "Why Gödel Didn't Have Church's Thesis," *Information and Control*, 54, pp. 3-24.

Dawson W. (1984) "The Reception of Gödel's Incompleteness Theorems," *Philosophy of Science Association*, 2, pp. 253-71, repr. in Shanker (1988) pp. 74-95.

Dawson W. (1997) *Logical Dilemmas: The Life and Work of Kurt Gödel*, Peters, Wellesley, MA.

Debray R. (1980) *Le Scribe: Genèse du politique*, Grasset, Paris.

Debray R. (1981) *Critique de la raison politique*, Gallimard, Paris.

Detlefsen M. (1979) "On Interpreting Gödel's Second Theorem," *Journal of Philosophical Logic*, 8, pp. 297-313, repr. in Shanker (1988) pp. 131-54.

Dowling W.F. (1989) "There Are No Safe Virus Tests," *American Mathematical Monthly*, 96, pp. 835-6.

Dummett M. (1959) "Wittgenstein's Philosophy of Mathematics," in Benacerraf and Putnam (1964) pp. 491-509.

Dunn J.M. (1979) "A Theorem in 3-Valued Model Theory with Connections to Number Theory, Type Theory and Relevant Logic," *Studia Logica*, 38, pp. 149-69.

Dunn J.M. (1986) "Relevance Logic and Entailment," in Gabbay and Guenthner (1983-9) vol. III, pp. 117-224.

Eagleton T. (1996) *The Illusions of Postmodernism*, Blackwell, Oxford.

Feferman S. (1960) "Arithmetization of Metamathematics in a General Setting," *Fundamenta Mathematicae*, 49, pp. 35-92.

Feferman S. (1962) "Transfinite Recursive Progressions of Axiomatic Theories," *Journal of Symbolic Logic*, 27, pp. 383-90.

Feferman S. (1983) "Kurt Gödel: Conviction and Caution," in Shanker (1988) pp. 96-114.

Feferman S. (1995) "Penrose's Gödelian Argument: A Review of Shadows of the Mind by Roger Penrose," *Psyche*, 2 (7), http://psyche.cs.monash.edu.au/v2/psyche-2-07-feferman.html.

Floyd J. (2001) "Prose versus Proof: Wittgenstein on Gödel, Tarski and Truth," *Philosophia Mathematica*, 9, pp. 280-307.

Floyd J., Putnam H. (2000) "A Note on Wittgenstein's 'Notorious Paragraph' about the Gödel Theorem," *Journal of Philosophy*, 97, pp. 624-32.

Fraassen B. van (1968) "Presuppositions, Implication and Self Reference," *Journal of Philosophy*, 65, pp. 136-51.

Fraenkel A., Bar-Hillel Y., Levy A. (1973) *Foundations of Set Theory*, North-Holland, Amsterdam.

Franzén T. (2005) *Gödel's Theorem: An Incomplete Guide to Its Use and Abuse*, Peters, Wellesley, MA.

Frege G. (1879) *Begriffsschrift, eine der arithmetischen nachgebildete Formelsprache des reinen Denkens*, Nebert, Halle.

Frege G. (1884) *Die Grundlagen der Arithmetik: Eine logisch-mathematische Untersuchung über den Begriff der Zahl*, Koebner, Breslau.

Frege G. (1903) *Grundgesetze der Arithmetik: Begriffsschriftlich abgeleitet*, vols I-II, Pohle, Jena.

Frixione M., Palladino D. (2004) *Funzioni, macchine, algoritmi: Introduzione alla teoria della computabilità*, Carocci, Rome.

Gabbay D., Guenthner F. (eds) (1983-9) *Handbook of Philosophical Logic*, vols I–IV, Kluwer, Dordrecht.

Gödel K. (1930) "Die Vollständigkeit der Axiome des logischen Funktionenkalküls," *Monatshefte für Mathematik und Physik*, 37, tr. "The Completeness of the Axioms of the Functional Calculus of Logic," in van Heijenoort (1967) pp. 582-91.

Gödel K. (1931) "Über formal unentscheidbare Sätze der *Principia Mathematica* und verwandter Systeme I," *Monatshefte für Mathematik und Physik*, 38, pp. 173-98, tr. "On Formally Undecidable Propositions of *Principia mathematica* and Related Systems I," in Shanker (1988) pp. 17-50.

Gödel K. (1944) "Russell's Mathematical Logic," in *The Philosophy of Bertrand Russell*, ed. P.A. Schlipp, Northwestern UP, Evanston, IL, pp. 125-53, repr. in Jacquette (2001) pp. 21-34.

Gödel K. (1947) "What is Cantor's Continuum Problem?," *American Mathematical Monthly*, 54, pp. 515-25.

Gödel K. (1986) *Collected Works I: Publications 1929-1936*, ed. S. Feferman et al., Oxford UP, Oxford.

Gödel K. (1990) *Collected Works II: Publications 1938-1974*, ed. S. Feferman et al., Oxford UP, Oxford.

Gödel K. (1995) *Collected Works III: Unpublished Essays and Lectures I*, ed. S. Feferman et al., Oxford UP, Oxford.

Goldstein R. (2005) *Incompleteness: The Proof and Paradox of Kurt Gödel*, Atlas, Norton, New York.

Hallett M. (1984) *Cantorian Set Theory and Limitation of Size*, Clarendon, Oxford.

Heijenoort J. van (ed.) (1967) *From Frege to Gödel: A Source Book in Mathematical Logic*, Harvard UP, Harvard.

Helmer O. (1937) "Perelman versus Gödel," *Mind*, 46, pp. 58-60.

Heyting A. (ed.) (1959) *Constructivity in Mathematics*, North Holland, Amsterdam.

Hilbert D. (1904) "Über die Grundlegung der Logik und der Arithmetik," in *Verhandlungen des Dritten Internationalen Mathematiker-Kongress in Heidelberg vom 8. bis 13. August 1904*, Teubner, Leipzig, pp. 174-85, tr. "On the Foundations of Logic and Arithmetic," in van Heijenoort (1967) pp. 129-38.

Hilbert D. (1925) "Über das Unendliche," *Jahresbericht der Deutschen Mathematiker-Vereinigung*, 36, s. I, pp. 201-15, tr. "On the Infinite," in van Heijenoort (1967) pp. 367-92.

Hilbert D., Bernays P. (1939) *Grundlagen der Mathematik*, Springer, Berlin.

Hintikka J. (1999) "Ludwig Wittgenstein: Half Truths and One-and-a-Half Truths," in idem, *Selected Papers*, vol. I, Kluwer, Dordrecht.

Hofstadter D. (1979) *Gödel, Escher, Bach: An Eternal Golden Braid*, Basic, New York.

Jacquette D. (ed.) (2001) *Philosophy of Logic: An Anthology*, Blackwell, Oxford.

Kadvany J. (1989) "Reflections on the Legacy of Kurt Gödel: Mathematics, Skepticism, Postmodernism," *The Philosophical Forum*, 20, pp. 161–81.

Kalmàr L. (1959) "An Argument Against the Plausibility of Church's Thesis," in Heyting (1959).

Kielkopf C. (1970) *Strict Finitism: An Examination of Ludwig Wittgenstein's Remarks on the Foundations of Mathematics*, Mouton, The Hague.

Kirkham R.L. (1992) *Theories of Truth: A Critical Introduction*, MIT Press, Cambridge, NJ.

Kleene S. (1952) *Introduction to Metamathematics*, North-Holland, Amsterdam.

Kleene S. (1976) "The Work of Kurt Gödel," *Journal of Symbolic Logic*, 41, pp. 761–8, repr. in Shanker (1988) pp. 48–71.

Kleene S. (1986) "Introductory Note to 1930b, 1931 and 1932b," in Gödel (1986) pp. 126–41.

Kreisel G. (1958) "Review of Wittgenstein's 'Remarks on the Foundations of Mathematics'," *British Journal for the Philosophy of Science*, 9, pp. 135–58.

Lindström P. (2001) "Penrose's New Argument," *Journal of Philosophical Logic*, 30, pp. 241–50.

Lolli G. (1994) *Incompletezza: Saggio su Kurt Gödel*, Il Mulino, Bologna.

Lolli G. (2002) *Filosofia della matematica: L'eredità del Novecento*, Il Mulino, Bologna.

Lolli G. (2004) *Da Euclide a Gödel*, Il Mulino, Bologna.

Lucas J.R. (1961) "Minds, Machines, and Gödel," *Philosophy*, 36, pp. 112–27, repr. in Anderson (1964) pp. 43–59.

Lucas J.R. (1996) "Minds, Machines, and Gödel: A Retrospect," in Millikan and Clark (1996) pp. 103–24.

Marconi D. (1984) "Wittgenstein on Contradiction and the Philosophy of Paraconsistent Logics," *History of Philosophy Quarterly*, 1, pp. 333–52.

Meyer R.K., Mortensen C. (1984) "Inconsistent Models for Relevant Arithmetic," *Journal of Symbolic Logic*, 49, pp. 917–29.

Millikan P.J.R., Clark A. (eds) (1996) *Machines and Thought: The Legacy of Alan Turing*, vol. I, Oxford UP, Oxford.

Moriconi E. (2001) *L'incompletezza dell'aritmetica*, in Bellotti, Moriconi, and Tesconi (2001) pp. 175–256.

Mortensen C. (1995) *Inconsistent Mathematics*, Kluwer, Dordrecht.

Nagel E., Newman J.R. (1958) *Gödel's Proof*, new edn ed. D. Hofstadter, New York UP, New York, 2001.

Palladino D. (2004) *Logica e Teorie formalizzate: Completezza, incompletezza, indecidibilità*, Carocci, Rome.

Penrose R. (1989) *The Emperor's New Mind: Concerning Computers, Minds and the Laws of Physics*, Oxford UP, Oxford.

Penrose R. (1994) *Shadows of the Mind: A Search for the Missing Science of Consciousness*, Oxford UP, Oxford.

Penrose R. (1996) "Beyond the Doubting of a Shadow: A Reply to Commentaries on *Shadows of the Mind*," *Psyche*, 2 (23), http://psyche.cs.monash.edu.au/v2/psyche-2-23-penrose.html.

Perelman C. (1936) "L'Antinomie de M. Gödel," *Académie Royale de Belgique, Bulletin de la Classe des Sciences*, Serie 5, 22, pp. 730-3.

Plebani M. (2007) *Wittgenstein e Gödel*, degree thesis.

Potter M. (2004) *Set Theory and Its Philosophy*, Oxford UP, Oxford.

Priest G. (1979) "The Logic of Paradox," *Journal of Philosophical Logic*, 8, pp. 219-41.

Priest G. (1984) "Logic of Paradox Revisited," *Journal of Philosophical Logic*, 13, pp. 153-79.

Priest G. (1987) *In Contradiction: A Study of the Transconsistent*, Nijhoff, Dordrecht, 2nd edn Oxford UP, Oxford, 2006.

Priest G. (1994) "Is Arithmetic Consistent?," *Mind*, 103, pp. 337-49.

Priest G. (1995) *Beyond the Limits of Thought*, Cambridge UP, Cambridge, 2nd edn Oxford UP, Oxford, 2002.

Priest G., Routley R., Norman J. (eds) (1989) *Paraconsistent Logic: Essays on the Inconsistent*, Philosophia, München.

Putnam H. (1961) "Minds and Machines," in S. Hook ed. *Dimensions of Mind*, Collier, New York, repr. in Anderson (1964) pp. 72-97.

Quine W.V.O. (1966) *The Ways of Paradox and Other Essays*, Random House, New York.

Quine W.V.O. (1970) *Philosophy of Logic*, Prentice Hall, Englewood Cliffs, NJ.

Ramsey F.P. (1931) *The Foundations of Mathematics and Other Logical Essays*, Routledge & Kegan Paul, London.

Resnik M.D. (1974) "On the Philosophical Significance of Consistency Proofs," *Journal of Symbolic Logic*, 3, pp. 133-47, repr. in Shanker (1988) pp. 115-30.

Rodych V. (1999) "Wittgenstein's Inversion of Gödel's Theorem," *Erkenntnis*, 51, pp. 173-206.

Rodych V. (2002) "Wittgenstein on Gödel: The Newly Published Remarks," *Erkenntnis*, 56, pp. 379-97.

Rodych V. (2003) "Misunderstanding Gödel: New Arguments about Wittgenstein and New Remarks by Wittgenstein," *Dialectica*, 57, pp. 279-313.

Rosser B. (1936) "Extension of Some Theorems of Gödel and Church,"*Journal of Symbolic Logic*, 1, pp. 87-91.

Rosser B. (1942) "The Burali-Forti Paradox," *Journal of Symbolic Logic*, 7, pp. 251-76.

Routley R. (1979) "Dialectical Logic, Semantics and Metamathematics," *Erkenntnis*, 14, pp. 301-31.

Russell B. (1903) *The Principles of Mathematics*, Cambridge UP, Cambridge.

Russell B., Whitehead A.N. (1910-13) *Principia Mathematica*, Cambridge UP, Cambridge.

Sainsbury R.M. (1995) *Paradoxes*, Cambridge UP, Cambridge.

Serres M. (ed.) (1989) *Eléments d'histoire des sciences*, Bordas, Paris.

Shanker S.G. (ed.) (1988) *Gödel's Theorem in Focus*, Croom Helm, London.

Shapiro S. (1998) "Incompleteness, Mechanism, and Optimism," *The Bulletin of Symbolic Logic*, 4, pp. 273-302.

Shapiro S. (2002) "Incompleteness and Inconsistency," *Mind*, 111, pp. 817-32.

Shapiro S. (2003) "Mechanism, Truth, and Penrose's New Argument," *Journal of Philosophical Logic*, 32, pp. 19-42.

Singh S. (1997) *Fermat's Last Theorem*, Fourth Estate, London.

Smullyan R. (1988) *Forever Undecided: A Puzzle Guide to Gödel*, Oxford UP, Oxford.

Smullyan R. (1992) *Gödel's Incompleteness Theorems*, Oxford UP, Oxford.

Sokal A., Bricmont J. (1997) *Impostures intellectuelles*, Jacob, Paris.

Sorensen R. (2003) *A Brief History of the Paradox: Philosophy and the Labyrinths of the Mind*, Oxford UP, Oxford.

Tarca L.V. (2007) "*Reductio Reductionis*: Sul significato filosofico della prova di Gödel," unpublished MS.

Tarski A. (1936) "O ugruntowaniu naukowej semantyki," *Przeglad Filozoficzny*, 39, pp. 50-7, tr. "The Establishment of Scientific Semantics," in Tarski (1956) pp. 401-8.

Tarski A. (1956) *Logic, Semantics, Metamathematics: Papers from 1923 to 1938*, Oxford UP, Oxford.

Turing A.M. (1937) "On Computable Numbers, with an Application to the Entscheidungsproblem," *Proceedings of the London Mathematical Society*, 42, pp. 230-65, repr. in Davis (1965) pp. 116-54.

Wittgenstein L. (1921) *Logisch-philosophische Abhandlung*, "Annalen der Naturphilosophie," 14, rev. edn *Tractatus logico-philosophicus*, Routledge & Kegan Paul, London, 1922.

Wittgenstein L. (1956) *Bemerkungen über die Grundlagen der Mathematik*, Blackwell, Oxford.

Wittgenstein L. (1964) *Philosophische Bemerkungen*, Blackwell, Oxford.

Wittgenstein L. (1967) *Wittgenstein und der Wiener Kreis (1929-32)*, Blackwell, Oxford.

Wittgenstein L. (1969) *Philosophische Grammatik*, Basil Blackwell, Oxford.

Wittgenstein L. (1976) *Lectures on the Foundations of Mathematics*, Cornell UP, Ithaca, NY.

Wittgenstein L. (2000) *The Big Typescript*, Springer, Wien.

Woods, J. (2003) *Paradox and Paraconsistency: Conflict Resolution in the Abstract Sciences*, Cambridge UP, Cambridge.

Zermelo E. (1908) "Untersuchungen über die Grundlagen der Mengenlehre I," *Mathematische Annalen*, 65, pp. 261–81, tr. "Investigations on the Foundations of Set Theory I," in van Heijenoort (1967) pp. 199–215.

Index

Printed and bound by CPI Group (UK) Ltd, Croydon, CR0 4YY

27/10/2024

14580383-0003